Yasir Nawab and Khubab Shaker (Eds.)
Textile Engineering

Also of interest

Particle Technology and Textiles.
Review of Applications
Edited by Jean Cornier, Franz Pursche, 2023
ISBN 978-3-11-067076-9, e-ISBN 978-3-11-067077-6

Encyclopedia of Color, Dyes, Pigments
Edited by Gerhard Pfaff, 2022
Volume 1 Antraquinonoid Pigments – Color Fundamentals
ISBN 978-3-11-058588-9, e-ISBN (PDF) 978-3-11-058807-1
Volume 2 Color Measurement – Metal Effect Pigments
ISBN 978-3-11-058684-8, e-ISBN (PDF) 978-3-11-058710-4
Volume 3 Mixed Metal Oxide Pigments – Zinc Sulfide Pigments
ISBN 978-3-11-058686-2, e-ISBN (PDF) 978-3-11-058712-8

Polymer Surface Characterization
Edited by Luigia Sabbatini, Elvira De Giglio, 2022
ISBN 978-3-11-070104-3, e-ISBN (PDF) 978-3-11-070114-2

Intelligent Materials and Structures
Haim Abramovich, 2021
ISBN 978-3-11-072669-5, e-ISBN 978-3-11-072670-1

Textile Chemistry
Thomas Bechtold, Tung Pham, geplant für 2023
ISBN 978-3-11-079569-1, e-ISBN 978-3-11-079573-8

Textile Engineering

An Introduction

2nd Edition

Edited by
Yasir Nawab and Khubab Shaker

DE GRUYTER
OLDENBOURG

Editors
Dr. Yasir Nawab
National Textile University
Sheikhupura Road
Faisalabad 37610
Pakistan
ynawab@ntu.edu.pk

Dr. Khubab Shaker
National Textile University
Sheikhupura Road
Faisalabad 37610
Pakistan
shaker.khubab@gmail.com

ISBN 978-3-11-079932-3
e-ISBN (PDF) 978-3-11-079941-5
e-ISBN (EPUB) 978-3-11-079948-4

Library of Congress Control Number: 2023931040

Bibliographic information published by the Deutsche Nationalbibliothek
The Deutsche Nationalbibliothek lists this publication in the Deutsche Nationalbibliografie;
detailed bibliographic data are available on the Internet at http://dnb.dnb.de.

© 2023 Walter de Gruyter GmbH, Berlin/Boston
Cover image: Obencem/iStock/Getty Images Plus
Typesetting: Integra Software Services Pvt. Ltd.
Printing and binding: CPI books GmbH, Leck

www.degruyter.com

Contents

Munir Ashraf

1 Introduction

Abstract: This chapter provides a brief introduction, history, and the evolution of textiles. It also covers the textile exports in the world and the status of the textile industry of Pakistan.

Keywords: Textile sector, history, exports of textiles

1.1 Introduction to Textiles

Textiles play a crucial role in human civilization to fulfill some basic human needs such as giving protection by clothes, tents, umbrellas, or shelter through a canopy. Textiles are also known as potential artistic materials and are widely used for artistic expression. The intriguing thing about textiles is that they are probably present in several household items such as towels, bed sheets, and clothes. In ancient times, humans wore clothes that are made of animal skins, grasses, and leaves. The desire for better clothing and apparel led to the growth of textile fiber production and the manufacturing process in the textile sector. Textile fibers played a crucial role to develop comfortable and sustainable physical structures of modern society. A fiber typically seems like hair because it has hundreds of times more length than its width. Both synthetic and natural materials can be used to make textile fibers. In terms of comfort and fashion, natural textile fibers are primarily suitable for human consumption [1]. Wool, silk, cotton, and jute are examples of natural fibers while spandex, polyester, rayon, acrylic, and nylon are known as synthetic fibers. The fibers are spun into strands to make textiles and then woven to make fabric. The textile sector is one of the biggest sectors in the world.

1.2 What Is Textile?

The word "textile" is derived from the Latin word "Textilis" meaning woven, fabric, or cloth. The product that is made from the interlacing of fibers and thread is referred to as textiles. Any intermediate or finished textile product obtained from any method falls within this broad definition. Thus, the term "textile" refers to the filament, yarn, fibers, knitted, braided, and nonwoven as well.

Several processes are involved in the manufacturing of textile products to form them wearable, and the names of these processes are spinning, knitting, weaving, processing, and garments manufacturing. Figure 1.1 depicts the flowchart of these different processes. A brief description of the manufacturing process is provided further.

https://doi.org/10.1515/9783110799415-001

1.2.1 Spinning

The spinning is the process of creating long, continuous strands of fibers (called yarns) by twisting together shorter strands of fibers. The process involves three basic steps: preparing the fibers, drafting, and spinning. The first step in the spinning process is to prepare the fibers. This involves cleaning and carding the fibers to remove any impurities and align the fibers so they are parallel. The next step is drafting, which involves pulling the fibers into long, thin strands that are more even in thickness. This is done by passing the fibers through a set of rollers or hand-pulling them. The final step in the spinning process is spinning, which involves twisting the drafted fibers together to create a single, continuous strand of yarn. This is done by either using a spinning wheel or a spindle, which rotates the fibers as they are twisted together. After the yarn is spun, it can be dyed, woven, or knitted into fabric or other textile products [1].

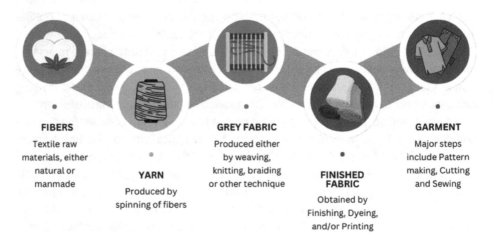

FIBERS

Textile raw materials, either natural or manmade

YARN

Produced by spinning of fibers

GREY FABRIC

Produced either by weaving, knitting, braiding or other technique

FINISHED FABRIC

Obtained by Finishing, Dyeing, and/or Printing

GARMENT

Major steps include Pattern making, Cutting and Sewing

Fig. 1.1: Flowchart of textile manufacturing.

1.3 Weaving and Knitting

Weaving and knitting are two fabric manufacturing processes. In weaving, two types of mutually perpendicular threads are used. The ones along the length of fabric are called warp threads, whereas the ones across the length of fabric are called weft threads. The fabric manufacturing takes place due to interlacement of warp and weft threads. The warp threads have to bear a lot of stresses and friction during weaving; therefore, a protective film called size is applied on them to prevent their breakage. The woven fabrics have normally high dimensional stability, i.e., they try to maintain their dimensions.

The second type of fabric manufacturing process is knitting in which the fabric is formed by the intermeshing of loops of yarns. There are two types of knitting processes: weft knitting and warp knitting. The knitted structures have normally less dimensional stability [2].

1.3.1 Coloration and Finishing

To add aesthetic and functional properties, both woven and knitted fabrics passed several processes in which they are treated with different chemicals to remove natural as well as added impurities. Once all the impurities are removed and the fabrics are completely white, then they segregated depending upon the requirements of customers. They are finished in white, dyed, or printed and then finished. The finishing is one of the most important processes in which functional properties such as water repellency, crease recovery, flame retardancy, and softness are imparted to textiles. These are chemical and water-intensive processes which generate a significant environmental impact [3].

1.3.2 Garment Manufacturing

The last step is transforming finished fabric into garment or home textile. Various steps are involved in the production of garment or home textile by garment manufacturing industries. Sampling, costing, designing, cutting, sewing, finishing, washing, packing, final inspection, dispatch, and many other steps are among them [4].

1.4 History of Textiles

Textile history is almost as old as human civilization. The earliest of Veda, Rigveda, consists of literary information regarding textile and thus called "weaving." Ramayana and Mahabharata, as prominent Indians, depict the wide range of fabrics within ancient India. It is suggested by archaeological evidence that humans used plants for weaving into garments and baskets about 23,000 years ago. After that, there was rapid advancement and diversification associated with fabric technology. Figure 1.2 illustrates the summary of the evolution of textile industry throughout the human history.

During the Industrial Revolution, weaving was mechanized which enabled fast production of cost-effective cloth. Until the nineteenth century, artificial fibers would not be a prominent category but emerged with the discovery of "Rayon," also known as artificial silk or semisynthetic fiber invented within the 1800s. Synthetic fibers

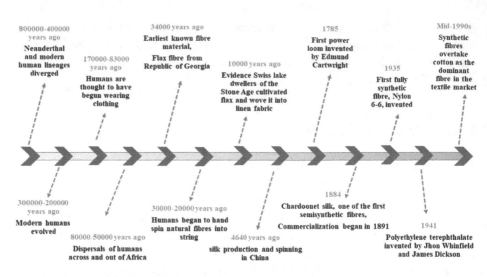

Fig. 1.2: Schematic of the history of the textile industry [5].

have revolutionized the textile industry since their inception and the history of the textile is represented in Tab. 1.1 [6].

Tab. 1.1: Detailed history timeline of textile [6].

Prehistoric and ancient times	
6300 BC	Turkish archaeologists discover finely woven fabric.
3000	In America, Pakistan, and western India, cotton was being cultivated.
2700	Silkworms were grown by the Chinese and sophisticated looms were designed to weave silk.
2500	Egyptians discovered the linen.
327	Alexander was impressed with gorgeous, printed cotton being manufactured in India.
300	Ancient Greeks and Romans established a massive textile business.
75	Silk had become the luxurious fabric by the Romans.
63	In Rome, awnings were made of cotton.

Middle Ages	
AD 768	At Lyons, silk industry was established by Charlemagne and brought wool from England.
900	Alfred promoted the growth of the wool industry in England.
1120	The first guild for woolen clothing was supported by Henry.
1153	First annual fabric fair organized by England.
1200	The spinning spindle was employed frequently.
1305	In Venice, more people engaged in the weaving wool.

Modern times	
1533	European spinning and weaving were inferior to Peruvian.
1589	William Lee had developed the machine for knitting hosiery.

Tab. 1.1 (continued)

Modern times

Early 1600	The dyeing and finishing of fabric were improved by textile workers in the Netherlands.
1631	Indian fine calico was imported by the Dutch East India Company.
1654	It was prohibited for English textile workers to travel to America.
1661	A power loom was constructed by a Danzig, Poland. He was drowned by the government, and the loom was destroyed.
1667	All individuals were required to be buried in woolen cloth under English law. More fabric was produced than what could be sold.
1669	In America, it was prohibited for the English colonies to exchange wool products
1696	Irish weavers manufactured fabric at a lower cost than English weavers. The weavers attempted to be silenced. Irish linen was the best of all.
1733	Flying shuttle loom was invented by an Englishman named John Kay.
1764	James Hargreaves designed the very first machine named pinning Jenny to simultaneously spin more than one strand of yarn.
1768	In America, contests in spinning and weaving were conducted to protest the Stamp Act.
1769	Richard Arkwright invented the "Water frame" a spinning machine that was powered by water.
1779	The water frame and the spinning jenny were components of the machine known as spinning mule which was invented by Samuel Crompton.
1785	The first power loom was patented by Edmund Cartwright.
1790	Samuel Slater invented the water-powered cotton spinning machines in America
1793	Cotton gin was developed by Eli Whitney.
1800	Woven line around 25 million yard was exported by Japan.
1804	Punched cards were used by Jacquard loom so that single weaver can produce complex patterned cloths.
1816	Many power looms were starting to be implemented in the industries in America.
1861	Machine-made uniforms were used by Union forces. Confederate uniforms were still spun and woven by hand.
1884	The first synthetic fiber (a kind of rayon) was manufactured by Hilaire Chardonnet.

The twentieth century

1900	Spinning and weaving were successfully transferred from domestic workshops to factories and mills as a result of the Industrial Revolution.
1910	The rayon fabric invented by Chardonnet was initially manufactured in America under the term of "artificial silk."
1935	Nylon was developed by Wallace C. Carrothers.
1940–1950	Artificial fibers like acrylic, polyester, and others were manufactured.
1960	Polyester fiber that is double-knit was introduced. The Fiber Product Identification Act for textiles was also passed into law.
1970	Computer-controlled knitting machines created textiles with incredibly complex patterns at breakneck speed.
Early 1980	The textile industry introduced robots.
Late 1980	Instead of using a single-shuttle, high-speed looms were utilized by textile mills that had several tiny shuttles known as darts. Other looms used no shuttles at all to weave.

At the first millennium's end, there were considerable changes in the processes such as weaving and spinning. Cotton was first introduced in Central Europe in the mid of fourteenth century. With the rapid rise in world population, drastic variations are needed by developed processes for meeting the needs or requirements. It originates from industrialization. The era of postindustrialization in textiles processes, raw materials, and machinery shows continuous innovation and improvement. Significant developments have been observed around the mid of twentieth century in raw materials, such as the preparation of polyester, polyacrylonitrile, and polyamides, and machinery such as spinning (open end) and weaving water jet loops took place. This innovation process is still at pace [1].

Moreover, printing and dyeing textile processes also show roots within the prehistoric era. The first solid proof of silk dyeing and brocades from social and religious records establishes that Indians were well aware of the process of dying in 2500 BC. It is thought in 3500 BC, the Chinese practiced dyeing; however, no solid evidence supports it. In 2500 BC, yellow and red colors for textile dyeing were obtained using safflower. Similarly, the whole color range for textiles was produced by Egyptians in 1450 BC. After the Roman Empire collapsed, there was no supporting evidence for textile development until 1371, when dyeing information was made public by dyers, who made their independent dye houses within Florence. Multiple shades were produced over 2,000 years ago by blending existing dyes to generate a broad color gamut. Table 1.2 illustrates the major improvements in printing and dyeing at various ages.

Tab. 1.2: History of dyeing material and different processes [7].

Time period	History of dyeing material and different techniques
3500 BC	In China, dyeing was practiced; however, solid proof is missing.
2500 BC	Social and religious records indicate that Indians used to dye brocades and silk.
715 BC	In Rome, wool dyeing is recognized as craft.
327 BC	In India, Alexander the Great refers "beautiful, printed cottons."
55 BC	The Romans discovered "picti" (painted individuals) dying themselves with woad in Gaul.
Second and third AD centuries	Indigo and madder dyed textiles were found on Roman graves that replaced the ancient Imperial Purple
AD 273	Emperor Aurelian prohibited his spouse from purchasing silk item that had been dyed with purpura. It was as expensive as gold.
AD 700s	Wax-resist method for dyeing was mentioned by a Chinese manuscript
AD 925	In Germany, Wool Dyers' Guilds first started
AD 1188	In London, the first occurrence of Guilds for the Dyers

Tab. 1.2 (continued)

Time period	History of dyeing material and different techniques
AD 1197	King John persuaded Parliament to regulate woolen clothing dyeing in order to shield the people from substandard products.
AD 1212	More than 200 dyers, tailors, and fullers could be found in Florence. Additionally, a directory of spinners and weavers was published.
AD 1290	In Germany, Woad, the only blue dye available at the time, has extensive applications in different fields.
AD 1321	Brazilwood was initially referenced as a dye, which comes from India and East Indies.
AD 1472	Company of London, the dyers was incorporated by Edward IV.
AD 1507	France, Germany, and Holland have started to growing dye plants as a sector.
AD 1614	In England, dyeing fabric "in the wood" was first presented: fustic, logwood, etc.
AD 1689	In Germany, calico printworks was first introduced, and they subsequently expanded into a significant industry.
AD 1745	In England, Indigo has begun to be cultivated, after the emerging trend, when it is more inexpensive to import from the East Indies.
AD 1774	Scheele, a Swedish chemist discovered that chlorine degraded vegetable colors by noticing cork in the hydrochloric acid vessel.
AD 1774	Sulfuric acid and Prussian Blue can be purchased commercially. In general, one of the earliest chemical dyes.
AD 1785	In England, Bell developed roller printing who earlier invented printing from plates.
AD 1786	In France, Bertholet recommended chlorinated water to bleach commercially. Other oxidizing substances including hydrogen peroxide, sodium perborate, and sodium peroxide started to be employed as well.
AD 1788	Picric acid (yellow disinfectant and dye) might be used to dye wool from acid dyebath.
AD 1790	Development of acid discharge by mordant printing.
AD 1856	First synthetic dye was discovered by William Henry Perkin.
AD 1858	Diazotization was discovered by Griess and coupling in or on the fiber.
AD 1858–1900	Invention of different methods to develop dyes.
AD 1914	90% dyestuffs were imported by USA.
AD 1922	The American Association of Textile Chemists and Colorists established its first subcommittee to investigate the wash fastness of both printed and dyed cotton, develop testing protocols, and establish criteria of fastness.

1.5 Status of Textiles in World Exports

Growth Rate of World GDP and Export

Fig. 1.3: World merchandise exports [8].

Top Textile Exporters in 2021 (By value)

	Exporters	Value of exports ($bn)	Growth rate (2020-2021)	Market shares
1	China	145.6	−5.5%	41.1%
2	European Union	73.6	13.7%	20.8%
3	India	22.2	47.8%	6.3%
4	Turkey	15.2	29.6%	4.3%
5	USA	13.1	15.3%	3.7%
6	Vietnam	11.5	17.1%	3.2%
7	Pakistan	9.2	29.2%	2.6%
8	South Korea	8.7	12.1%	2.5%
9	Taiwan	8.6	21.3%	2.4%
10	Japan	6.2	10.6%	1.8%

Fig. 1.4: World export of Textile [8].

Based on the World Trade Statistical Review (2022) released by the World trade organization (WTO), world apparel exports surged in 2021 as the global economy recovered from COVID, although global textile exports increased very slowly as a result of a large trade volume the year before. In 2021, World clothing exports entirely recovered to the pre-COVID level and exceeded $548.8 billion, a significant rise of 21.9% from 2020. Comparatively, the value of global textile exports grew more slowly in 2021 at

7.8%, lagging behind many other industries. This pattern was understandable because the textile industry remained strong in 2020 due to the significant increase in the demand for personal protective equipment throughout the pandemic. According to WTO, the growth of global merchandise trade will be reduced to 3.5% in 2022 and to only 1% in 2023. Thus, the global textile and apparel trade is anticipated to experience sluggish growth or a slight drop in the next couple of years [8]. The data of growth rate of world GDP and exports during the past three years is shown in Figure 1.3.

In 2021, China, European Union, and India continued to be the top three textile exporters in the world and continuing a trend that has persisted for more than ten years. These top three countries produced 68% of all textile exports globally in 2021, which is similar to the 66.9% in 2018–2019 before to the pandemic. However, the ten largest textiles exporters experienced a wide range of growth rates in 2021, from -5.5% (China) to 47.8% (India) [8]. Figure 1.4 and 1.5 shows the export values, growth rate, and the market shares of the top ten textile and clothing exporters in 2021. The global export patterns of clothing were considerably disrupted in 2021 by the surge in consumer demand and COVID-related supply chain disruptions. In 2021, China, Bangladesh, Turkey, and India experienced a growth of over 20% in their clothing exports. China witnessed a

Top Clothing Exporters in 2021 (By value)

	Exporters	Value of exports ($bn)	Growth rate (2020-2021)	Market shares
1	China	176.1	24.4%	32.1%
2	European Union	151.0	19.8%	27.5%
3	Bangladesh	35.8	30.4%	6.5%
4	Vietnam	31.2	11.1%	5.7%
5	Turkey	18.7	22.0%	3.4%
6	India	16.2	24.5%	2.9%
7	Malaysia	14.5	46.8%	2.6%
8	Indonesia	9.4	23.9%	1.7%
9	Hong Kong	8.6	4.3%	1.6%
10	Pakistan	8.5	36.9%	1.5%

Fig. 1.5: World export of Clothing [8].

growth of 24%, while Bangladesh and India reported a surge of 30% and 24%, respectively. Similarly, the clothing exports of Turkey is escalated by 22%.

1.6 Status of Textile in Pakistan

Currently, the textile industry of Pakistan is an organized sector at a large scale along with a highly fragmented small-scale sector and cottage. The organized sector includes integrated textile mills with huge spinning units and small units of shuttle-less looms. Downstream industries (Hosiery, Towels, Garment, Finishing, and Weaving) exhibit greater export potential as part of an unorganized sector. Few units have grown well toward International Scale and progressive with business philosophy. The spinning sector of Pakistan in June 2021 is composed of textile units of about 517 (spinning units (477) and composite units (40)). The number of conventional looms and shuttle-less looms includes 375,000 and 28,500. Rise in the spinning sector with the demand for export and production of cotton followed by processing and weaving sector. Weaving units based on air jets are either set as independent units or in collaboration with processing or spinning units.

1.6.1 Growth of Textile Industry

In Pakistan's textile industry, the loom capacity and spindle capacity remained unchanged at 9,084 and 13.41 million, respectively, during 2020–2021. There is an increase in production of yarn from 3.049 billion kg to 3.441 billion kg in 2019–2020 and

Tab. 1.3: Growth in capacity and production of textile industry in Pakistan [9].

	2017–2018	2018–2019	2019–2020	2020–2021
Capacity				
Spindles	13.410	13.409	13.409	13.409
Rotors	198,801	198,801	198,801	198,801
Looms (mill sector)	9,804	9,804	9,084	9,084
Shuttle less	28,500	28,500	28,500	28,500
Power looms	**375,000**	375,000	375,000	375,000
Looms total	**412,583**	**412,583**	**412,583**	**412,583**
Production				
Production of yarn (M kg)	3,430.1	3,431.2	3,049.6	3,441.6
Production of cloth (M m^2)				
Mill sector	1,043.7	1,046.0	931.0	1,048.4
Nonmill sector	8,127.2	8,128.8	7,226.6	8,128.8
Total	**9,170.9**	**9,147.8**	**8,157.6**	**9,177.39**

2020–2021, respectively. The detail of the growth in capacity and production is given in Tab. 1.3.

1.6.2 Growth in Textiles Exports

The Pakistan textile industry provides a major contribution to the export earning of Pakistan. The export basket includes a wide range of products including cotton fibers, yarn, yarn derived from materials other than cotton, fabrics, towels, bed sheets, canvas, carpets, tents, and a variety of clothing. Due to its intrinsic expertise in the global market for its conventional goods, Pakistan's textile industry has the potential to perform better in terms of manufacturing and export. However, a significant investment in modern technology and machinery infrastructure is required to preserve its position and move in high value-added goods for the market share that has risen in value. The immediate areas of attention for any organization should be employees' training, increased labor productivity, research and development, product diversity, and branding. The detail of the export performance of Pakistan textile sector in different years is presented in Tab. 1.4.

Tab. 1.4: The export performance of Pakistan textile industry [9].

Products	2016–2017	2017–2018	2018–2019	2019–2020	2020–2021
Cotton and cotton textile	12,205	13,220	13,031	12,212	15,030
Synthetic fabrics	204	310	298	315	370
Wool and carpets	79	76	67	54	74
Total textiles	12,531	13,606	13,396	12,581	15,474
All exports	20,448	23,222	22,979	21,394	25,304
Textile as % of total exports	61.3%	58.6%	58.3%	58.8%	61.2%

Textile industry is the most important industrial sector in Pakistan with the largest production chain and the possibility for value addition at every stage of the process from cotton through ginning, spinning, fabric, dyeing and finishing, and clothing. The industry delivers around one-fourth of all industrial value-added, employment has given to approximately 40% of all industrial laborers, uses about 40% of all manufacturing-related bank credit, and represents 8% of GDP. Apart from periodic and cyclical swings, textile exports have consistently accounted for roughly 54% of total national exports. Pakistan is the third-largest consumer of cotton and the fourth-largest producer in the worlds, although its comparative advantage is decreasing because of the export of low-value textile goods.

References

[1]   T. Gries, D. Veit, and B. Wulfhorst, *Textile technology: an introduction*. Carl Hanser Verlag GmbH Co KG, 2015.

[2]   S. Adanur, *Handbook of weaving*. CRC press, 2020.

[3]   T. A. Khattab, M. S. Abdelrahman, and M. Rehan, "Textile dyeing industry: Environmental impacts and remediation," Environmental Science and Pollution Research, vol. 27, pp. 3803–3818, 2020.

[4]   T. Karthik, P. Ganesan, and D. Gopalakrishnan, *Apparel Manufacturing Technology*. CRC Press, 2016.

[5]   D. Sanders, A. Grunden, and R. R. Dunn, "A review of clothing microbiology: The history of clothing and the role of microbes in textiles," Biology Letters, vol. 17, no. 1, p. 20200700, 2021.

[6]   Utah Education Network Team, "Textile history timeline," UTAH EDUCATION NETWORK. [Online]. Available: https://www.uen.org/cte/family/clothing-2/downloads/textiles/timeline.pdf. [Accessed: 23-Feb-2023].

[7]   H. Ben Slama, et al., "Diversity of Synthetic Dyes from Textile Industries, Discharge Impacts and Treatment Methods," Applied Science, vol. 11, no. 14, p. 6255, Jul. 2021.

[8]   Dr Sheng Lu, "World Textiles and Clothing Trade in 2021: A Statistical Review," JustStyle, 2022. [Online]. Available: https://www.just-style.com/analysis/world-textiles-and-clothing-trade-in-2021-a-statistical-review/.

[9]   M. F. Khan, "Performance of Textile industry," Textile Commissioner's Organization. [Online]. Available: https://www.tco.com.pk/documents/7098418c64.pdf. [Accessed: 23-Feb-2023].

Madeha Jabbar, Khubab Shaker
2 Textile Raw Materials

Abstract: The textile raw materials are the fibers that can be converted into yarns and fabric of any desired specifications and requirements. A fiber is defined as a delicate, hair-like portion of the tissues of a plant, animal, or other substance; that is very small in diameter as compared to its length. The suitability of fiber for the textile application depends mainly on its length, strength, uniformity, fineness, and elongation. These properties enable the fibers to be twisted together to form a yarn. A higher fiber length allows the fibers to be twisted easily into the yarn, while strength and elongation help to withstand the tensions during the subsequent textiles processes. The fiber uniformity and fineness help to produce a yarn of uniform cross-section. The textile fibers are either nature-based (obtained from plants or animals) or man-made.

Keywords: plant fibers, animal fibers, man-made fibers, synthetic fibers

2.1 Classification

Textile raw materials are fibers that can be converted into yarns and fabric of any desired specifications and requirements. A fiber is defined as a delicate, hair-like portion of the tissues of a plant, animal, or other substance that is very small in diameter as compared to its length. The classification of textile raw materials on the basis of origin is shown in Fig. 2.1.

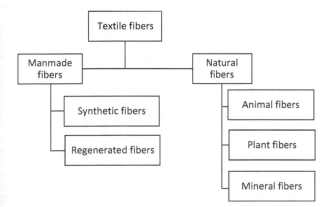

Fig. 2.1: Textile fiber classification on the basis of origin.

https://doi.org/10.1515/9783110799415-002

The natural fibers are those provided by nature in readymade form and need to be extracted only. On the other hand, man-made fibers are generated by humans from things that were not in fiber form previously [1].

2.2 Natural Fibers

Natural fibers such as cotton, flax, jute, silk, and wool have been in considerable demand for textiles for ages due to their renewability, eco-friendly nature, and ease of availability. In addition, the natural fibers have low density, better mechanical and thermal properties, and are biodegradable. The natural fibers are categorized into

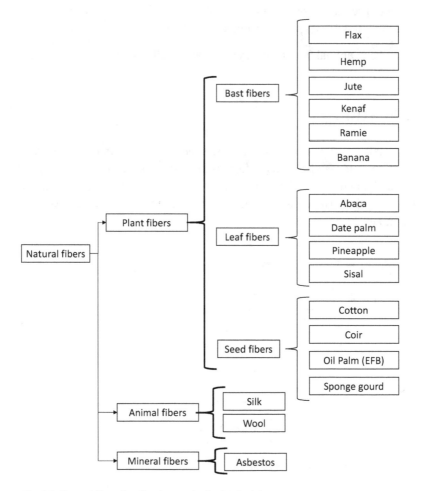

Fig. 2.2: Natural fiber classification on the basis of origin.

three main classes depending on the nature of the source (origin), that is, plant, animal, and mineral fibers as shown in Fig. 2.2.

Plant fibers (also termed as vegetable fibers) include the most important of the entire textile fibers "cotton" together with flax, jute, hemp, sisal, and other fibers obtained from plants. All these fibers are mainly composed of varying fractions of cellulose, hemicellulose, and lignin. The generalized structure of a plant fiber is shown in Fig. 2.3. Cellulose is the basic constituent for all lignocellulosic fibers, contributing about 50–70% of the fiber, and therefore determines the properties of the fiber. Hemicellulose acts as cementing material, joining cellulosic microfibrils, while lignin improves hydrophobic properties by water proofing of cell wall [2]. The chemical composition of some common plant fibers is given in Tab. 2.1. The natural fibers are collected from different portions of plants and hence classified on this basis into stem/bast, leaf, seed fibers, and so on.

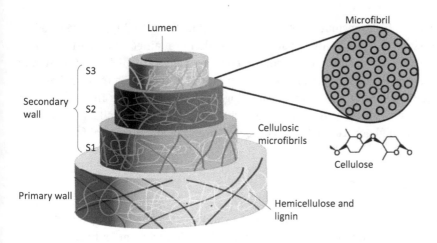

Fig. 2.3: Generalized structure of a plant fiber [3].

Tab. 2.1: Chemical composition of some common plant fibers [4, 5].

Fiber	Lignin (%)	Cellulose (%)	Hemicellulose (%)	Density (g/cm³)
Bast fibers				
Flax	1–5	60–85	14–20.6	1.4–1.5
Hemp	2–10	55–90	12–22.4	1.4–1.6
Jute	9–26	45–72	13.6–24	1.3–1.5
Kenaf	8–22	31–72	18.5–29.7	1.4–1.5
Ramie	0.4–9.1	61.8–91	5.3–16.7	1.4–1.5
Banana	5–10	60–85	6–8	1.35

Tab. 2.1 (continued)

Fiber	Lignin (%)	Cellulose (%)	Hemicellulose (%)	Density (g/cm³)
Leaf fibers				
Abaca	5–15.1	53–68	17.5–25	1.5
Date palm	27.7	33.9	26.1	0.7–1.2
Pineapple	5–12	70–82	17.5	0.8–1.6
Sisal	8–14	52.8–78	10–19.3	1.3–1.5
Agave	15	80	3–6	–
Fruit/seed fibers				
Cotton	0.75	82.7–98	3–5.7	1.5–1.6
Coir	40–45.8	32–43.8	0.15–20	1.15–1.3
Oil palm EFB	4–20	30–65	15–38.7	0.7–1.55
Sponge gourd	11.2–16.4	59.4–66.6	16.5–29.5	0.72–0.92

The animal fibers include wool and other hair like fibers and fibers produced by silkworm (such as silk). It also includes the feathers obtained from certain birds. These animal fibers are protein-based, the complex material which most of animal body is made of. Mineral fibers are of less importance in the textile trade. Asbestos is most useful fiber of this class. The outstanding property of asbestos fiber is its resistance to heat and burning. They are also highly resistant to acids, alkalis, and other chemicals. These fibers are used to make special fireproof and industrial fabrics.

2.2.1 Cotton Fibers

Cotton is a soft, staple fiber that grows in a form known as boll (protective capsule) around the seeds of cotton plant. Cotton is generally obtained from four species in the genus *Gossypium*, that is, arboretum, hirsutum, barbadense, and herbaceum. The fiber is either spun into yarn to produce breathable textiles by weaving or knitting or directly converted to web by nonwoven technique. These textiles are the most widely used form of textile for clothing. The value of natural fibers produced globally in 2018 was estimated to be $60 billion, and the shares of cotton fiber were about 66%. China, India, the USA, Brazil, and Pakistan are the leading cotton producer countries [6].

Cotton picking by hand is still practical in nearly all countries. The common practice in hand picking is to pick the seed cotton and the boll and put it into a sack. An experienced adult can pick 300 lbs. of seed cotton per day under normal conditions. In case of automatic picking, two types of pickers are used, namely stripper and spindle picker. Hand picking is advantageous to machine picking, as fibers are picked only from the completely mature capsules. After picking, the cotton is taken to the ginning factory where fibers are separated from the seeds. The beaters used for ginning are either saw gin or roller gin. Saw gin is more economical due to advanced automation and mechanization. The seeds separated are used to extract oil for edibles or soap/candle manufacture. A pound of seed cotton can be obtained from 50 to 100 bolls depending on the nature of plant and condition under which it is grown.

The cotton fiber is classified on the basis of fineness, staple length, maturity, degree of contamination, and strength. The fineness of fiber is denoted in dtex, that is, number of grams per 1,000 m. Another approach to denote fineness of cotton fiber is in terms of micronaire value, that is, number of micrograms per inch. The staple length of fiber is critical to determine the fineness of yarn produced from this fiber. A high staple length of cotton fiber produces a finer yarn. Fiber maturity affects the shade of fiber. Immature and weak cotton have a cloudy appearance, while mature cotton appears bright and has a thick cell wall.

Under a light microscope, cotton fibers are recognized by the presence of the lumen and convolutions, that is, twists along the length of the fiber. Another unique feature of cotton fibers is the reversal in the direction of the spiral (fibril) structure or helix along the length of the fiber. The important characteristics of cotton fiber are given in Tab. 2.2.

Tab. 2.2: Properties of cotton fiber.

Parameter	Value
Fineness	1–4 dtex/2.3–6.9 micronaire
Fiber length	10–60 mm
Moisture regain	8.5%
Breaking strength	25–50 cN/tex
Elongation	5–10%
Color	Creamy yellow

Cotton fiber turns yellow at temperatures above 110 °C. It is not damaged by sunlight; however gradual loss of strength occurs on longer exposure to sunlight. Being cellulosic in nature, it dissolves in the concentrated solution of acids, but has excellent resistant to alkalis. A strong caustic solution causes the fibers to swell. Fungus and bacteria attack and degrade cotton. It contains mineral nutrients (salts of Na, K, and Mg and Ca) and starch which promote growth of fungus and mildew, especially in highly humid conditions. Bacteria and fungus discharge enzymes which attack the cellulose and convert it to sugar. For protection, cotton is treated with materials which either inhibit the growth or kill these microorganisms.

Cotton has a low thermal conductivity and is therefore suitable for both summer clothing (preventing skin from heat) and winter clothing (preserving warmth of body). It is widely preferred for apparels including dresses, shirts, trousers, jeans, blouses, skirts, suits, dresses, active wear, swimwear, and hosiery. Home textile articles made of cotton include curtains, draperies, bedspreads, sheets, towels, tablecloths, table mats, and napkins. Industrial and technical applications of cotton fiber include tents, tarpaulins, ropes, bags, shoes, filter cloth, and medical supplies. Sometimes, the cotton fiber is blended with different fibers to get a single yarn, with desired properties. Cotton blends easily with other fibers; mostly with the polyester and viscose.

2.2.2 Flax

Flax is probably the oldest textile fiber known to mankind. The fiber is obtained from the stem of a plant *Linum usitatissimum*, which is 80–120 cm high. The flax-woven fabric is also called linen. The flax plant is thought to have arrived in Europe with the first farmers, and in the Stone Age people were usually dressed in linen clothes [7]. The linen fabric used to wrap the mummies found in Egyptian tombs is approximated to be more than 3,000 years old. But even long before that time flax was being used for various applications.

The two variants of flax plant include fiber flax and seed flax. The fiber flax is optimized for the production of thin strong fibers, while seed flax gives far more linseed and coarser fibers. The fiber flax grows in humid, moderate areas, while oil flax grows in dry, warm areas. The characteristics of flax differ depending on the sowing and growing conditions, affecting stem length, thickness, and the number of branching. The cross breeding these two types had resulted in the cultivation of oil-fiber linen named as combination linen. Flax grows in moderate climates and is presently cultivated in large parts of Western and Eastern Europe, Canada, the USA, and Russia.

The harvesting of flax plant is done by pulling the stalk either by hand or using a mechanical puller. Sometimes harvesting is done by cutting the plants close to the ground but pulling is preferred in order to retain the longest fiber length. The flax stalk bundles are then allowed to dry. Rippling is the next process, resulting in the removal of flower heads (pods) and leaves from the stem. These stems are then spread over the ground and allowed for retting. Retting breaks down the pectin layer, holding the fiber bundles together in stem by the combined action of bacteria and moisture.

The types of retting commonly employed for flax are water retting, enzyme retting, and dew retting [3]. In water retting, the bundles of stem are immersed in the running water (rivers) or standing water (ponds or specially prepared pits). The anaerobic bacteria cause the fermentation, thus degrading the pectin and other binder substances. The enzyme retting employs the use of warm water and enzymes to degrade the pectin. It is a controlled method, preferred for the production of very fine fibers, but it is rather laborious process. In dew retting, the flax stems are spread over the field. The humidity in environment causes the growth of indigenous aerobic fungi which partly degrades the stem. Dew retting is inexpensive process, taking about three to seven weeks depending on the weather conditions.

The retted stems having fibers loosened from the stem are dried and the stem is broken by passing it between the fluted rollers. The broken stem parts are removed from the fiber bundles in the scutching process. The scutching machine consists of two interpenetrating rollers equipped with three or more knives. The knives scrape along the fiber to remove the wooden stem. The scutched fiber bundles are still relatively coarse, thick, and ribbon-shaped. After scutching, the fibers are combed (hackling process), producing a thinner finer with circular fiber structure. Properties of flax fiber are given in Tab. 2.3.

Tab. 2.3: Properties of flax fiber.

Parameter	Value
Fineness	10–40 dtex
Diameter	10–80 µm
Fiber length	200–800 mm
Moisture regain	12%
Breaking strength	30–55 cN/tex
Elongation	2–3%

The flax fiber has the ability to absorb a lot of moisture and dry quickly, keeping the wearer dry and cool. Therefore, it is preferred in the manufacturing of apparels for summer after cotton. The linen has very low elasticity and shows better dimensional stability. It has vast uses such as suits, shirting, bed linen, table wear, kitchen towels, upholstery, surgical thread, sewing thread, artist's canvases, high quality papers, luggage fabrics, paneling, filtration, reinforced plastics, and composite materials. The ability of flax fiber to absorb water rapidly is particularly useful in towel trade.

2.2.3 Jute

Jute is known as the "golden fiber" due to its golden-brown color and is obtained from plants in the genus *Corchorus*, family Malvaceae. Jute belongs to bast fiber category and is normally spun in the form of coarse threads. In the plant stem, jute fibers surround the woody core and are embedded in the nonfibrous material under the bark. The fibrous strands closer to the bark run the full length of stem while the strands farther from bark become progressively short. Jute fiber consists of overlapping cells, about 0.1 in. long. Contrary to most vegetable fibers which consist mainly of cellulose, jute fibers are part cellulose and part lignin.

Cultivation requires well-drained, fertile soil, and a hot, moist climate. The jute plant becomes ready for harvesting in 120 days and is either pulled by hand or cut by sharp edge. These stems are then tied into bundles for retting. The process of retting involves immersion of the stems in water until the bacterial action releases the fibers within the stalk. It takes about 12–25 days for completion of retting. Stripping is done for removal of jute fibers from stem. The most common method is manual stripping, performed by beating the bark gently with a wooden mallet, starting from stem base. The fibers are then separated and dried. The properties of jute fiber are given in Tab. 2.4.

Tab. 2.4: Properties of jute fiber.

Parameter	Value
Fineness	2–3 dtex
Diameter	15–25 μm
Fiber length	650–750 mm
Moisture regain	13.75%
Breaking strength	30–34 cN/tex
Elongation	2–8.2%

Jute is a strong and cheap fiber, having good insulating properties for both thermal and acoustic applications. The current annual worldwide production of jute fiber is about 3.2 million tons and used for various end uses, for example, hessian sacks, garden twine, ropes, and carpets. In agriculture sector, jute is a popular choice to control soil erosion, seed protection, and weed control. It is used for technical applications in the area of geotextiles. Green polymeric composites are also produced using jute as a reinforcement material [8–10]. The jute is being replaced by synthetic materials for many of these uses, but the biodegradation and sustainability are main advantages of jute over synthetic fibers.

2.2.4 Other Plant Fibers

Other commonly used plant fibers include hemp, kenaf, ramie, sisal, pineapple, abaca, date palm, agave, coir, and empty fruit bunch [2].

The stalk of hemp plant produces two types of fibers: long (bast) fibers and short (core) fibers. Bast fibers can be cleaned, spun, and then woven or knitted into many fabrics suitable for durable and comfortable clothing and housewares. Fabrics produced from blended yarns with at least 50% hemp content help to block the UV rays. Hemp fibers are longer, strong, more lustrous, absorbent, and mildew resistant as compared to the cotton fiber. Hemp textiles are extremely versatile and used in the production of apparel, shoes, rugs, canvas, and upholstery.

Ramie belongs to the category of bast fibers and is one of the oldest vegetable fibers used for mummy cloths in Egypt. It needs chemical treatment to remove the gums and pectin found in the bark. The fiber is very fine like and being naturally white in color does not need bleaching. It is extremely absorbent fiber and naturally resistant to stains and bacteria, mildew, or insect attack. On the other hand, it is low in elasticity and lacks resiliency, low abrasion resistance, and wrinkles easily [11]. Ramie is commonly used in clothing, tablecloths, upholstery fabrics, napkins, and so on, while the technical applications include sewing thread, canvas, packing materials, filter cloth, and fishing nets.

Banana plant is not only a source of delicious fruit but also provides fiber for textile applications. The fiber is obtained after the fruit is harvested. The small pieces of banana plant trunk are put through a softening process for mechanical extraction of the fibers with subsequent bleaching and drying. In the recent past, banana fiber had a very limited application for making items like mats, ropes, and some composite materials. With the increasing demand for eco-friendly fabrics, it is finding applications in other fields such as apparels and home furnishings.

Hibiscus cannabinus, also known as kenaf, is a tall and slender plant, resembling bamboo or jute. The plant helps in maintaining a sustainable environment by absorbing huge quantities of CO_2 (approximately three times more than a tree). It can grow up to 14 feet and yields around 6–10 tons of fiber per acre. The fibers are obtained from bark (40%) and core (60%) of the plant [12]. Products made of kenaf fiber include ropes, animal bedding, paper products, woven or knitted fabrics, and polymer matrix composites. A comparison of the properties of common natural fibers is provided in Tab. 2.5.

Tab. 2.5: Properties of other bast fibers.

Parameter	Hemp	Ramie	Banana	Kenaf
Fineness	2–6 dtex	5–13 dtex	120–150 dtex	30–35 dtex
Diameter	15–50 μm	40–80 μm	80–250 μm	20–25 μm
Fiber length	600–750 mm	500 mm	340–850 mm	1–7 mm
Moisture regain	12%	8.5%	9–9.6%	9–10%
Breaking strength	35–70 cN/tex	40–70 cN/tex	46–64 cN/tex	25–30 cN/tex
Elongation	1–6%	2–3%	1.4–2.6%	1.3–5.5%

Sisal fiber is derived from the leaves of the sisal plant. It is usually obtained by machine decortications in which the leaf is crushed between rollers and then mechanically scraped. The fiber is then washed and dried by mechanical or natural means. The dried fiber represents only 4% of the total weight of the leaf. Once it is dried the fiber is mechanically double-brushed. The lustrous strand of sisal fiber, usually creamy white in color is relatively coarse and inflexible. It is valued for cordage (ropes, baler, binder twines, etc.) owing to its high strength. The higher grade fiber after treatment is converted into yarns and used by the carpet industry.

Pineapple is mainly cultivated for fruit [13], but its leaves are also a source of textile fibers. The fiber extraction from leaves is done by a combination of two techniques, scrapping followed by water retting. Scrapping involves scratching the layer of leaf and crushing to facilitate the entry of microbes for retting process. These crushed leaves are water-retted, and fibers are mechanically segregated, rinsed with water, and hang-dried in air. Date palm plant is another source of fibers, providing fibers from four different parts, leaf fibers, bast fibers from stem, wood fibers from trunk, and surface fibers around the trunk.

Coir is a short and coarse fiber obtained from the outer shell of coconut. Its low decomposition rate is a key advantage for making durable geotextiles. The coir fiber has highest lignin content (more than 40%) as compared to other plant fibers, making it a strong fiber [14]. Coir fiber is not a preferred choice for clothing but commonly used for doormats, brushes, rugs, insulation panels, and sacking for packaging. Oil palm plant is grown globally for edible oil. However, the trunk, frond, and empty fruit bunch (EFB) of oil palm tree can be used for the extraction of lignocellulosic fibers. The EFB can yield up to 73% fibers, offering additional advantages of low cost and ease of availability. Properties of these leaf and seed/fruit fibers are given in Tab. 2.6.

Tab. 2.6: Properties of other leaf and seed/fruit fibers.

Parameter	Pineapple	Sisal	EFB	Coir
Diameter	20–80 μm	22–80 μm	150–500 μm	12–25 μm
Fiber length	300–500 mm	1,000–1,250 mm	30 mm	150–300 mm
Moisture regain	11.8%	11%	8–10%	13%
Breaking strength	11–45 cN/tex	30–45 cN/tex	50–400 MPa	12–18 cN/tex
Elongation	1.6–3%	2–3%	3.2–30%	25–27%

2.2.5 Wool

Wool is an animal fiber obtained by shearing the fibrous covering of sheep and is produced in almost all parts of the world. Sheep are commonly shaved for their fleece once or twice a year and the raw wool obtained is known as fleece. An efficient shearer would remove the fleece from a sheep in 2 min. Wool is also removed from the pelts of slaughtered sheep by chemical treatment or bacterial action without damaging the hide. Raw wool is often dirty and contaminated with natural fats, grease, and perspiration residues. All these impurities are removed during wool scouring and wool carbonizing to get cleaned wool.

The breed of the sheep as well as the environmental conditions strongly affects the quality of wool. The wool also differs in fineness, length, and purity depending on the body part of sheep from which it is taken. Wool may be broadly classified into fine wool, medium wool, long wool, and carpet wool. It is spun to produce two types of yarns, that is, woolen and worsted. Woolen yarns are usually made from short staple fibers which are held loosely and given only a limited twist during spinning. Worsted yarns are much finer, regular, tightly twisted, and smoother than woolen. These are usually spun from longer staple fibers.

The wool fiber has an inherently three-dimensional crimp due to its unique chemical and physical structure [15]. The products made from wool exhibit a rapid wrinkle recovery, abrasion resistance, bulk, and warmth. It is naturally elastic and resilient, having the ability to absorb up to 30% moisture before feeling damp. The wool fiber is rather flame-resistant inherently (does not support combustion) and also exhibits self-extinguishing properties, that is, stops burning when it is removed from the source of the flame.

2.2.6 Silk

Silk is a protein fiber of insect origin, being produced as a fine filament of long length from the body fluid of silkworm (*Bombyx mori*). The silkworms eat only the leaves of mulberry tree. The four stages in the life cycle of a silkworm are: egg, caterpillar, larva (cocoon), and butterfly. Caterpillars are produced from eggs after hatching for 12 days. During the growth period of caterpillar, fresh mulberry leaves are its food. After 35 days, the caterpillars are ready for spinning silk. They stop eating and produce their cocoon in a few days. Silkworm makes its cocoon from a twin filament that extrudes from two silk glands in its head. These filaments are coated and glued together by gummy substance called sericin. The worm gradually gets covered and captivated in a strongly structured cocoon made from continuous silk strand (may be up to a mile in length). This filament silk is unwound from the cocoons.

Silk fiber is considered as a semicrystalline fiber, having a crystallinity of 62–65% in when obtained from *B. mori* silkworm and 50–63% for wild origin silkworms. The fiber is a polypeptide, formed from mainly four different amino acids (glycine, alanine, serine, and tyrosine). Silk fibers display remarkable mechanical properties, being strong, extensible, mechanically compressible, and high thermal stability. Their mechanical performance depends on the size and orientation of crystalline domains and their connectivity to the less crystalline domains. The pattern of light-reflection of its triangular shape is partly responsible for the luster of silk fiber [16]. The properties of silk fiber are given in Tab. 2.7.

Tab. 2.7: Properties of silk fiber.

Parameter	Value
Fineness	1–3.5 dtex
Diameter	10–13 µm
Fiber length	700–1,500 m
Density	1.37 g/cm^3
Moisture regain	9–11%
Breaking strength	25–50 cN/tex
Elongation	10–25%
Color	Lustrous white

The fibroin of silk is decomposed by concentrated acids into the constituent amino acids. Silk is more resistant to alkalis and organic solvents, except hydrogen bond breaking solvents. Continuous exposure of silk fiber to sunlight results in strength loss. It begins to yellow at high temperatures and disintegrates above 165 °C. The moisture absorption results in a temporary 10–25% strength loss of silk fiber. It is easily attacked by moth and mildew.

2.3 Man-made Fibers

The man-made fibers are classified into synthetic and regenerated fibers as shown in Fig. 2.4. The polymers used for the spinning of synthetic fibers are chemical-based, while regenerated fibers are derived from a natural polymer, most commonly cellulose [17].

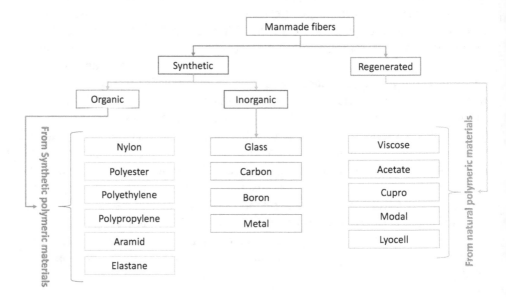

Fig. 2.4: Classification of man-made fibers.

2.3.1 Spinning of Man-made Fibers

There are three most common techniques employed in the production of man-made fibers namely wet-spinning, melt-spinning, and dry-spinning as shown in Fig. 2.5. These techniques vary in the method of liquefying the raw material (powder or pellet). The term spinning here defines the extrusion process of liquefied polymer through spinnerets and subsequent solidification during a continuous flow. The melt-spinning involves a simple transformation of the polymer physical state and is applied only to polymer having a melting temperature, for example, PA 6, PA 6.6, PES, and PP. In melt-spinning, the extruded polymer is transformed directly into a filament owing to its fast cooling, and cross-sectional form remains unchanged [18].

In case of solution spinning, the polymer is dissolved in variable concentrations according to the kind of polymer and of solvent to produce a viscous liquid (dope). It is used for the polymers that degrade thermally at a temperature lower than the melting point [cellulosic fibers, polyacrylonitrile (PAN)]. The extruded filaments are subject to structural changes due to solvent extraction from the polymer mass. The solution spinning is further divided into type, dry-spinning, and wet-spinning. In dry-spinning, the solvent is removed by flow of warm gas directed to the extruded filaments. The wet-spinning method involves introduction of extruded polymeric viscose into a coagulation bath, where water behaves as a solvent for the polymer solvent and as a nonsolvent for the polymer mass.

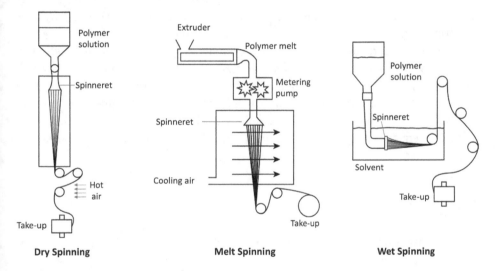

Fig. 2.5: Spinning techniques for man-made fibers.

2.4 Regenerated Fibers

The regenerated fibers are produced by dissolving the cellulose in a certain chemical and converting this solution again into the fiber form (by viscose method). Since these fibers consist of cellulose, they are also termed as "regenerated cellulose fibers." The cellulose has an empirical formula of $(C_6H_{10}O_5)_n$, where $(C_6H_{10}O_5)$ represents the glucose molecule and "n" is degree of polymerization, DP (number of glucose molecules constituting the cellulose). The α-cellulose is insoluble in cold dilute NaOH and has a DP greater than 200, while β-cellulose (hemicellulose soluble in cold dilute NaOH) has DP less than 200. The raw material for regenerated fibers is wood pulp, which is refined to get higher percentage of α-cellulose in the fiber. Rayon is the generic term used for the family of regenerated fibers and most commonly includes viscose, acetate, and cuprammonium (cupro or cupra) rayon. The regenerated fibers are produced according to the viscose-spinning method.

2.4.1 Viscose Fiber

The production of viscose fiber involves the process of solution spinning. The solution for spinning is prepared by treating cellulose with the NaOH, producing alkali cellulose. This alkali cellulose reacts with carbon disulfide to give cellulose xanthate, which on dissolution in NaOH gives viscose solution. This solution is extruded from the spinneret into the spinning bath. The composition of spinning bath, its temperature, and spinning speed is adjusted according to the type of fiber to be spun. The solidification of solution into yarn takes place in the spinning bath. These spun fibers are drawn to achieve a regular orientation of the chain molecules.

The spun fibers are further treated to remove impurities, increase brightness, and improve adhesive and frictional properties. These treatments may include washing, bleaching, and application of some finish. The spinning process parameters allow the manipulation of properties to a wider level, producing certain fibers like high tenacity viscose, highly crimped viscose, hollow fibers, and modal. By the addition of suitable chemicals in viscose solution or spinning bath, spin-dyed, flame-retardant fibers, and so on can be produced.

2.4.2 Acetate Fiber

The cellulose is mixed with acetic anhydride and glacial acetic acid under addition of wet-splitting chemicals. The spinning is carried out according to the dry-spinning technique. The spinning solution is extruded through spinneret, and filaments are passed through hot air stream in the quench duct. It results in the evaporation of solvent and solidification of filaments. During the passage, the filaments are also drawn, combined, oiled, and wound onto the bobbins. The properties of common regenerated fibers are given in Tab. 2.8.

2.5 Synthetic Fibers

The synthetic fibers are result of the extensive research to improve the properties of naturally occurring animal and plant fibers. The synthetic fibers are produced by the extrusion of a polymeric material having synthetic origin through spinneret into air

Tab. 2.8: Properties of regenerated fibers.

Parameter	Viscose	Cupro	Acetate
Fiber length	38–200 mm	–	40–120 mm
Tenacity	15–30 cN/tex	16–25 cN/tex	20–40 cN/tex
Elongation	15–30%	16–25%	20–40%
Density	1.52 g/cm^3	1.52 g/cm^3	1.29–1.33 g/cm^3
Modulus	8–12 cN/tex	40–60 cN/tex	8 cN/tex
Melting point	175–205 °C	175–205 °C	250 °C

or water. These fiber forming polymers are obtained generally from petrochemicals and are therefore termed as synthetic fibers.

2.5.1 Polyamides

Polyamide is a family of synthetic fibers, also termed as nylon family. Nylon 6 and Nylon 6.6 are the most important members of this family. Nylon 6.6 was the first synthetic fiber produced in 1935 in the USA. A parallel development in Germany led to the production of continuous filament (Nylon 6) in 1939 [19]. Nylon fibers are made up of linear macromolecules whose structural units are linked by the amide (–NH–CO–) group. Therefore, these fibers are termed as polyamides. The most common way for the production of nylon polymers is by the condensation of diamines with diacids. The molecules are directional in nylon 6, with all amide links in a particular direction, while a reversal in the order of alternate amide linkages is observed in nylon 6.6.

Nylon 6
–NH–CH$_2$–CH$_2$–CH$_2$–CH$_2$–CH$_2$–CO–
Nylon 6.6
–NH–CH$_2$–CH$_2$–CH$_2$–CH$_2$–CH$_2$– CH$_2$–NH –OC– CH$_2$–CH$_2$–CH$_2$–CH$_2$–CH$_2$–CO–

The nylon fibers are produced by the extrusion of molten polymer and no solvents are involved. The polymer is initially in the form of chips or pellets and is melted by a heated grid or by an extruder where screw (single or twin) forces the polymer along a heated tube. The molten polymer is fed to a controlled metering pump which helps to control the linear density of the fiber. The molten polymer is fed to the spinneret at 280–300 °C, against 50–70 MPa pressure. The spinneret is a stainless-steel plate (5 mm or more thick), having a number of small holes (each having diameter of 100–400 μm). These number of holes correspond to the number of filaments required in the final yarn. The polymer emerging from the spinneret holes is drawn down by a take-up reel and undergoes a considerable acceleration. The accelerated filaments solidify the cool air. The properties of nylon fiber are given in Tab. 2.9.

Tab. 2.9: Properties of nylon fiber.

Parameter	Nylon 6.6	Nylon 6
Melting point	255–260 °C	215–220 °C
Softening point	235 °C	170 °C
Moisture regain	3.5–4.5%	3.5–4.5%
Modulus	20–35 cN/tex	15–35 cN/tex
Breaking strength	40–60 cN/tex	40–60 cN/tex
Elongation	20–30%	20–40%

Polyamides lose strength and elongation when exposed to air at temperatures above 100 °C. The UV radiations also have a detrimental effect on the mechanical properties of polyamides. Nylons are slowly affected by the water at the boiling point. Nylon 6.6 is inert to alkali solution while sensitive to acids. Both Nylon 6 and 6.6 are inert to common organic solvents; however they dissolve in concentrated formic acid and phenols.

2.5.2 Polyester

Polyethylene terephthalate, also called polyester fiber, dominates the world synthetic fibers industry. It is a lightweight, strong, and wrinkle-resistant, having good wash and wear properties. Polyester is produced by the condensation polymerization of a dicarboxylic acid with a diol, containing in-chain ester units as polymer-forming chain linkage [20]:

$$\text{Alcohol}\,(R-OH) + \text{carboxylic acid}\,(R-COOH) \rightarrow \text{ester}$$

where R is an alkyl group.

The filament polyester fiber is produced by melt-spinning under different conditions. The polymer is melted in a screw-extruder and forced through tiny holes (0.180–0.400 mm) in the spinneret plate by displacement gear pump. The polymer solidifies as it emerges from the spinneret. The cooling process is accelerated by controlled flow of air. The polymer is also drawn down, that is, stretched in semimolten state to induce molecular order and orientation in the fiber. The common properties of polyester fiber are given in Tab. 2.10.

2.5.3 Other Organic Synthetic Fibers

Other commonly used synthetic fibers include polyolefins, aramids, acrylics, and elastane. Olefins are long-chain synthetic polymer-based fibers having at least 85 wt% of ethylene, propylene, or other olefin monomers. Polypropylene (PP) and polyethylene

Tab. 2.10: Properties of polyester fiber.

Parameter	Polyester
Melting point	480 °C
Softening point	460 °C
Modulus	800–1,000 cN/tex
Breaking strength	40–60 cN/tex
Elongation	10–20%
Density	1.22–1.38 g/cm^3
Moisture regain	0.4–0.8%

(PE) are the two most common members of the olefin family. PP fiber is not attacked by bacteria, other micro-organisms, or insects. It is also hydrophobic and inherently resistant to the growth of mildew and mold. PE fiber has low density and low melting point. When density of PE is enhanced, strong fibers like HDPE and UHMPE are obtained.

Aramids are also called aromatic amides and are available in two types: meta and para-aramids. The meta-aramids are resistant to high temperatures, while para-aramids are high-modulus/high-strength fibers. However, para-aramid offer low resistance to strong acids and alkalis, and strength loss is observed when exposed to light or saturated steam. Meta-aramids are typically used in protective clothing (firefighter, auto racer, heat shields, etc.) and similar applications while para-aramids find application in ballistic protection, cut-resistant apparel/gloves, as reinforcement for polymer composites, and so on.

Elastane (also called spandex or Lycra) is a long-chain polyurethane, also known as a polyurea copolymer. It is a lightweight, synthetic fiber invented in 1958 by DuPont. Elastance is commonly used for stretchable clothing such as sportswear, body suits, gloves, tights, belts, and socks. A comparison of the properties is given in Tab. 2.11.

Tab. 2.11: Properties of synthetic fibers.

Parameter	Polyethylene	Polypropylene	*p*-Aramid	Elastane
Tenacity	32–65 cN/tex	15–60 cN/tex	19–25 g/den	4–12 cN/tex
Elongation	10–45%	15–200%	3.4%	400–800%
Density	0.95–0.96 g/cm^3	0.9 g/cm^3	1.15 g/cm^3	1.1–1.3 g/cm^3
Modulus	15–30 cN/tex	13–15 cN/tex	550–1,300 g/den	0.05–0.1 cN/tex
Melting point	125–135 °C	160–175 °C	–	230 °C

2.5.4 Inorganic Synthetic Fibers

Glass fiber and carbon fiber are the widely used examples of inorganic synthetic fibers. Glass is an inorganic fiber that does not support combustion and retains 25% of its strength at high temperature (540 °C). Silica sand, soda ash, and limestone are the main ingredients of glass fiber. The silica sand functions as the glass former, while limestone and soda ash help to lower the melting temperature. The low thermal expansion coefficient and low thermal conductivity make the glass fiber a dimensionally stable material as compared to organic synthetic fibers.

Preparation of glass fiber is a two-stage process, that is, formation of glass and formation of fibers. During the glass formation stage, mixing of raw materials is done in a batch furnace at 1,700 °C to get a homogeneous glass melt. This molten glass is fed to a series of heated platinum bushings, each having number of holes in its base. The glass flows under hydrostatic pressure and fine filaments having average diameter of 8–15 μm are mechanically drawn, as shown in Fig. 2.6. The fineness of drawn filaments is governed by diameter and length of holes in bushing, glass melt viscosity, hydrostatic pressure, and winding speed. The drawn fibers are instantly cooled, coated with a size material, collected into a strand, and wound on bobbin.

Glass fiber is inert to majority of chemicals except hydrofluoric, phosphoric acids, and strong alkalis. The blend of properties including high strength, low moisture

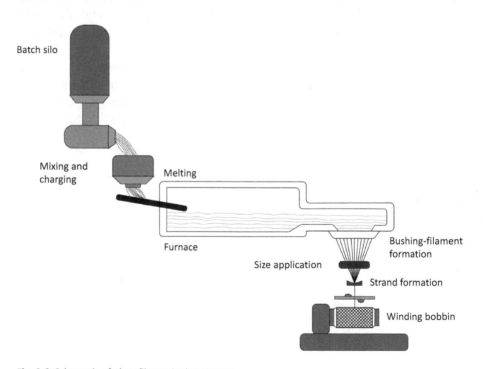

Fig. 2.6: Schematic of glass fiber-spinning process.

absorption, heat resistance, and low dielectric constant makes glass fiber an ideal reinforcement for printed circuit boards. The glass fiber is available in unidirectional or bidirectional, woven and nonwoven forms. These are extensively used as reinforcement for composite material in automotive market, civil construction, sporting goods, aviation and aerospace, boats and marine, electronics, home, and wind energy.

Carbon fiber is a long, thin strand composed mainly of carbon atoms, with an average diameter of 5–10 μm. These carbon atoms are bonded together and are more or less parallel to the fiber axis, making it incredibly strong. Carbon fibers were developed in the 1950s by heating strands of rayon until carbonization. It was an inefficient process, yielding about 20% carbon only that too had poor strength and stiffness. In theearly 1960s, carbon fiber was produced using PAN as a raw material. During the 1970s, a petroleum pitch derived from oil processing was introduced as an alternative material for the production of carbon fiber. It yielded about 85% carbon that had brilliant flexural strength, but poor compression strength, and were not widely accepted. Nowadays, approximately 90% of the carbon fiber is produced from PAN and remaining 10% from rayon or petroleum pitch.

A typical process used to form carbon fibers from PAN includes spinning, stabilization, carbonizing, surface treating, and sizing as shown in Figure 2.7. The acrylonitrile powder is mixed with another material like methyl acrylate or methyl methacrylate and is reacted with a catalyst to form PAN. It is then either spun into fibers via coagulation or heated and pumped into a chamber through tiny jets. The solvent is evaporated in the chamber to produce a solid fiber. The produced fibers are then washed and drawn to achieve desired fiber diameter. The drawn fibers are stabilized by heating in the presence of air at 200–300 °C for 30–120 min. As a result of stabilization process, fibers get oxygen from air and rearrange the atomic bonding pattern to produce a thermally stable ladder bonding.

The stabilized fibers are heated to a temperature of 1,000–3,000 °C for several minutes in a furnace filled with a mixture of gases other than oxygen (to prevent the fiber burning at very high temperature). Heated fibers lose their noncarbon atoms in the form of gases (carbon monoxide, carbon dioxide, ammonia, nitrogen, hydrogen, and others). The remaining carbon atoms form tightly bonded carbon crystals that are aligned parallel to the fiber longitudinal axis. The surface treatment of these fibers is performed to improve the fiber bonding properties. The carbon fibers are coated with epoxy, polyester, or other materials to protect from damage during subsequent processes. The coated carbon fibers are wound on bobbins for transportation.

The carbon fiber-reinforced composites are used in the automotive and aerospace industry, sports, and many other components where lightweight and high strength are needed. Carbon fibers have high electric conductivity and excellent EMI shielding property making it ideal for EMI shielding applications. Carbon fibers are dimensionally stable, sustain their mechanical performance at high temperatures, and have low thermal expansion coefficient.

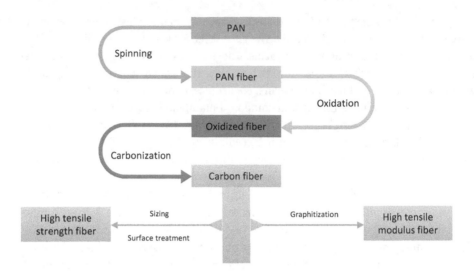

Fig. 2.7: Flowchart showing carbon fiber-spinning processes.

References

[1]   B. Wulfhorst, O. Maetschke, M. Osterloh, A. Busgen, and K.-P. Weber, *Textile Technology*. Germany: Hanser Publishers, 2006. doi: 10.1002/14356007.a26_553.

[2]   K. Shaker and Y. Nawab, *Lignocellulosic Fibers Sustainable Biomaterials for Green Composites*. Cham, Switzerland: Springer, 2022. doi: https://doi.org/10.1007/978-3-030-97413-8.

[3]   A. Ali, et al., "Hydrophobic treatment of natural fibers and their composites – A review," Journal of Industrial Textiles, vol. 47, no. 8, pp. 2153–2183, 2018. doi: 10.1177/1528083716654468.

[4]   E. Vázquez-Núñez, A. M. Avecilla-Ramírez, B. Vergara-Porras, and M. D. R. López-Cuellar, "Green composites and their contribution toward sustainability: A review," Polymers and Polymer Composites, vol. 29, no. 9, pp. S1588–S1608, 2021. doi: 10.1177/09673911211009372

[5]   K. Shaker, Y. Nawab, and M. Jabbar, "Bio-composites: Eco-friendly substitute of glass fiber composites," in *Handbook of Nanomaterials and Nanocomposites for Energy and Environmental Applications*, O. V. Kharissova, L. M. T. Martínez, and B. I. Kharisov, Eds. Springer, 2020. doi: https://doi.org/10.1007/978-3-030-11155-7_108-1.

[6]   T. Townsend, *World Natural Fibre Production and Employment*, vol. 1. Elsevier Ltd, 2020. doi: 10.1016/B978-0-12-818398-4.00002-5.

[7]   H. L. Bos, *The Potential of Flax Fibres as Reinforcement for Composite Materials*. Eindhoven, Netherlands: Eindhoven University of Technology, 2004.

[8]   M. M. Rehman, M. Zeeshan, K. Shaker, and Y. Nawab, "Effect of micro-crystalline cellulose particles on mechanical properties of alkaline treated jute fabric reinforced green epoxy composite," Cellulose, vol. 4, 2019. doi: 10.1007/s10570-019-02679-4.

[9]   M. H. Ameer, et al., "Interdependence of moisture, mechanical properties, and hydrophobic treatment of jute fibre-reinforced composite materials," The Journal of the Textile Institute, vol. 108, no. 10, pp. 1768–1776, 2017. doi: 10.1080/00405000.2017.1285201.

[10]   A. Ali, et al., "Experimental and numerical characterization of mechanical properties of carbon / jute fabric reinforced epoxy hybrid composites," Journal of Mechanical Science and Technology, vol. 33, no. 9, pp. 4217–4226, 2019. doi: 10.1007/s12206-019-0817-9.

[11]   B. J. Collier and P. G. Tortora, *Understanding Textiles*, 6th ed. New Jersey: Prentice Hall, 2001.

[12]   A. Kakoty, W. S. N. Sangma, A. R. Phukan, and B. B. Kalita, "Extraction of kenaf fiber and its physico-chemical properties for various end uses," International Journal of Chemical Studies, vol. 7, no. 3, pp. 2617–2620, 2019.

[13]   K. Ismoilov, S. Chauhan, M. Yang, and Q. Heng, "Spinning system for pineapple leaf fiber via cotton spinning system by solo and binary blending and identifying yarn properties," Journal of Textile Science and Technology, vol. 05, no. 04, 86–91, 2019. doi: 10.4236/jtst.2019.54008.

[14]   B. P. Corbman, *Textiles: Fiber to Fabric*, 6th ed. New York: McGraw-Hill, 1983.

[15]   M. M. Houck, Ed., *Identification of Textile Fibers*. Cambridge: Woodhead Publishing Limited, 2009.

[16]   R. M. Kozłowski, *Handbook of Natural Fibers – Volume 1: Types, Properties and Factors Affecting Breeding and Cultivation*, no. 118. Philadelphia, 2012.

[17]   Y. Nawab, T. Hamdani, and K. Shaker, Eds., *Structural Textile Design: Interlacing and Interlooping*. Boca Raton: CRC Press, 2017.

[18]   C. Andreoli and F. Freti, Eds., *Man-made Fibers*, 2nd ed. Italy: ACIMIT. 2006.

[19]   J. E. McIntyre, Ed., *Synthetic Fibres: Nylon, Polyester, Acrylic, Polyolefin*, 1st ed. Cambridge: Woodhead Publishing Limited, 2005.

[20]   A. R. Horrocks and S. C. Anand, Eds., *Handbook of Technical Textiles*. New York: Woodhead Publishing Ltd, 2000.

Zulfiqar Ali

3 Yarn Manufacturing

Abstract: This chapter is intended to provide a basic understanding of the yarn manufacturing process to the readers. Yarns are the basic building blocks of fabric. Therefore, an overview of yarn types and different yarn manufacturing techniques are necessary to explain the structure–property relationship of fabric. A brief introduction about process flowcharts for 100% cotton and cotton-blended yarns, preparatory, and spinning operations of yarns is part of this chapter.

Keywords: Cotton blends, yarn, spinning, roving, drawing, winding

3.1 Introduction

First a brief introduction is given of the terms used in yarn manufacturing.

3.1.1 Yarn

It is an assembly of substantial length and a relatively small cross-section of fibers and or/filaments with or without a twist. Yarn occurs in the following forms [1]:

(a) A number of fibers twisted together
(b) A number of filaments laid together without twist (a zero-twist yarn)
(c) A number of filaments laid together with a degree of twist
(d) A single filament with or without twist (a monofilament)
(e) A narrow strip of material, such as paper, plastic film, or metal foil, with or without twist, intended for use in textile construction

3.1.2 Spun Yarn

The yarn which consists of staple fibers held together by twist is known as spun yarn. The yarns produced on ring-spinning, open-end rotor spinning, and air-jet spinning systems; all are spun yarns.

3.1.3 Yarn Number

The number which shows the fineness or coarseness of yarn is called the yarn number. There are two systems of yarn numbering:

https://doi.org/10.1515/9783110799415-003

(i)   Indirect system
(ii) Direct system

In a direct system, yarn number is called the linear density of yarn with units of tex, denier, dTex, and so on. Similarly, in the indirect system, yarn number is called the yarn count with units of N_{EC}, N_m, N_{woolen}, and so on.
 The detail of these systems is as follows:

3.1.3.1 Indirect system

It is used for the measurement of length per unit weight of the yarn. In this weight is kept constant while length is a variable.
 In the indirect system, yarn thickness and yarn number are inversely proportional. It means that the yarn count increases as the yarn weight decreases and yarn becomes finer.
 The indirect system is also known as the English system of counting. The most used indirect numbering systems are

(1)   English cotton $N_{ec} = \dfrac{\text{No. of 840 hanks}}{\text{lbs}}$

(2)   Metric system $N_m = \dfrac{K_m}{K_g}$

(3)   Worsted $N = \dfrac{\text{No. of 560 hanks}}{\text{lb}}$

Different hank lengths for different fibers are given in the Table 3.1.

Tab. 3.1: Different hank lengths for different fibers.

Nature of fiber	Hank length
Cotton	840 yards
Spun silk	840 yards
Wool	256 yards
Worsted	560 yards
Linen	300 yards
Jute	14,400 yards

3.1.3.2 Direct system

It is used for the measurement of linear density which is the weight per unit length of yarn. In this system, yarn length is kept constant and weight is variable:

$$\text{Linear density} = \frac{\text{mass}}{\text{length}}$$

In a direct system, yarn thickness and yarn number are directly proportional. The most widely used direct numbering systems are

(a) Tex (Tex = no. of grams/1,000 m)
(b) Grex (Grex = no. of grams/10,000 m)
(c) Denier (Denier = no. of grams/9,000 m)

3.2 Yarn Production

Yarn production is a process of converting fibers into yarn. It consists of different processes. Flowcharts of 100% cotton and PC carded ring spun, 100% cotton and PC combed ring spun, open-end rotor spun, and air-jet spun yarns are given in Tables 3.2 to 3.7, respectively.

Tab. 3.2: Flowchart of 100% cotton-carded ring-spun yarn along with input and output.

Input materials	Process machines	Output materials
Raw cotton	Blow room	Lap/tufts
Lap	Card	Carded sliver
Carded sliver	Drawing 1	Breaker drawn sliver
Breaker sliver	Drawing 2	Finisher drawn sliver
Finisher drawn sliver	Simplex	Roving
Roving	Ring frame	Yarn
Yarn	Auto winding	Yarn cones

Tab. 3.3: Flowchart of PC-carded ring-spun yarn along with input and output.

Input materials	Process machines	Output materials
Polyester staple fiber	Blow room	Lap/tufts
Lap/tufts	Card	Carded sliver (polyester)
Carded sliver (polyester)	Predrawing	Predrawn polyester sliver
Raw cotton	*Blow room*	*Lap/tufts*
Lap/tufts	*Card*	*Carded sliver (cotton)*
Carded sliver (cotton) + predrawn polyester sliver	Drawing 1	Breaker drawn sliver (blended)
Breaker sliver (blended)	Drawing 2	Interdrawn sliver (blended)
Interdrawn sliver (blended)	Drawing 3	Finisher drawn sliver (blended)
Finisher drawn sliver (blended)	Simplex	PC-blended roving
PC-blended roving	Ring frame	PC-blended yarn
PC-blended yarn	Auto winding	PC-blended yarn cones

Tab. 3.4: Flowchart of 100% cotton combed ring-spun yarn along with input and output.

Input material	Process machines	Output materials
Raw cotton	Blow room	Lap/tufts
Lap	Card	Carded sliver
Carded sliver	Drawing	Drawn sliver
Drawn sliver	Lap former	Mini lap
Mini lap	Comber	Combed sliver
Combed sliver	Drawing	Comber drawn sliver
Comber drawn sliver	Simplex	Roving
Roving	Ring frame	Yarn
Yarn	Auto winding	Yarn cones

Tab. 3.5: Flowchart of PC combed ring-spun yarn along with input and output.

Input material	Process machines	Output materials
Polyester staple fiber	Blow room	Lap/tufts
Lap/tufts	Card	Carded sliver (polyester)
Carded sliver (polyester)	Predrawing	Predrawn polyester sliver
Raw cotton	*Blow room*	*Lap/tufts*
Lap	*Card*	*Carded sliver*
Carded sliver	*Drawing*	*Drawn sliver*
Drawn sliver	*Lap former*	*Mini lap*
Mini lap	*Comber*	*Combed sliver*
Combed sliver (cotton) + predrawn polyester sliver	Drawing 1	Breaker drawn combed sliver (blended)
Breaker drawn combed sliver (blended)	Drawing 2	Interdrawn combed sliver (blended)
Interdrawn combed sliver (blended)	Drawing 3	Finisher drawn combed sliver (blended)
Finisher drawn combed sliver (blended)	Simplex	PC-blended combed roving
PC-blended combed roving	Ring frame	PC-blended combed yarn
PC-blended combed yarn	Auto winding	PC-blended combed yarn cones

Tab. 3.6: Flow chart of cotton-carded rotor-spun yarn along with input and output.

Input material	Process machines	Output materials
Raw cotton	Blow room	Lap/tufts
Lap	Card	Carded sliver
Carded sliver	Drawing	Drawn sliver
Drawn sliver	Open-end rotor machine	Yarn cone

Tab. 3.7: Flowchart of air-jet spun yarn along with input and output.

Input material	Process machines	Output materials
Raw cotton	Blow room	Lap/tufts
Lap	Card	Carded sliver
Carded sliver	Drawing 1	Drawn sliver
Breaker sliver	Drawing 2	Finisher drawn sliver
Finisher drawn sliver	Air-jet machine	Yarn cone

3.2.1 Lap

A sheet of fibers wrapped around a rod/or roller to facilitate transfer from one process to the other is called a lap [2]. Output of scutcher and lap former are examples of the lap. The width of the lap produced by the lap former is about one-third of the scutcher lap so it is called a minilap.

3.2.2 Sliver

The assemblage of loose, roughly parallel fibers in continuous form without twist is called sliver [2]. Outputs of card and drawing are the examples of sliver.

3.2.3 Roving

A name given, individually or collectively, to the relatively fine fibrous strands used in the later or final processes of preparation for spinning is called roving [2]. Output of the roving frame which is used as input for the ring frame is an example of roving.

3.3 Basic Preparatory Processes for Spinning Operations

Fibers in the bale form are not suitable to start yarn manufacturing. There are a number of processes to make them suitable for spinning. The following subsections describe the basic preparatory processes which may be used as per the end-product requirements.

3.3.1 Preparation of Cotton/Polyester to Feed the Blow Room

After opening the strips of selected bale/bales and cleaning the sides, small tufts from the bales are taken and spread on the floor-selected area for making the layers of the heap as shown in Fig. 3.1. Several horizontal layers upon layers are made till the end of bales. This heap of cotton/polyester is left for 24 h to release the packing pressure and condition the material so that moisture in the material becomes homogeneous.

Fig. 3.1: Heap of opened cotton.

3.3.2 Blow Room for Cotton

Blow room line consists of different machines and each manufacturer provides its own line of machines. The sequence of machines in a typical cotton blow room line is shown in Fig. 3.2.

Objectives of a blow room line are as follows:
– Opening: To open the compressed fibers up to very small tufts.
– Cleaning: To remove the impurities like seed fragments, stem pieces, leaf particles, neps, short fibers, dust, and sand.
– Mixing and blending: To make homogenous mixture of the material.
– Dedusting: To extract the dust if present.
– Uniform feed for card: To convert the mass of fibers into a thick sheet called lap which should be uniform length and widthwise or to provide output in the form of tufts of optimum size.

The following is the blow room operation summary:

The cotton is fed manually on the feed belt of the blending feeder. The opening, cleaning, and blending are carried out by inclined lattice and evener roller. The stripper

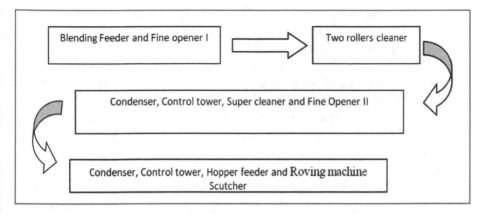

Fig. 3.2: Ohara blow room line.

roller transports this material to the feed lattice of fine opener-I. Figure 3.3 shows the material flow through the blending feeder machine. The beater of fine opener-I beats it against the grid bars for cleaning. This material is sucked by a condenser transport fan through two rollers cleaner, which performs the opening, cleaning, and dust extraction of cotton.

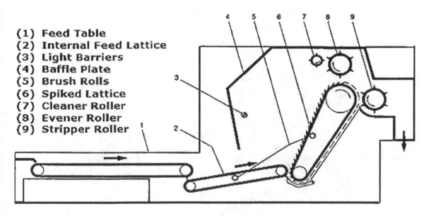

(1) **Feed Table**
(2) **Internal Feed Lattice**
(3) **Light Barriers**
(4) **Baffle Plate**
(5) **Brush Rolls**
(6) **Spiked Lattice**
(7) **Cleaner Roller**
(8) **Evener Roller**
(9) **Stripper Roller**

Fig. 3.3: Material flow through blending feeder.

The cage of the condenser separates the dusty air from the material and delivers it to the control tower feed rollers. A beater in the control tower beats the material against the grid bars and transfers to the spiked rollers of super cleaner. The six spiked rollers of super cleaner perform the opening and cleaning of material and deliver it to the feed lattice of fine opener-II. Where a Kirschner beater with steel pins further reduces the tuft size and helps in cleaning.

The opened and cleaned material from fine opener-II is sucked by the cage con-denser fan and then it is delivered to the control tower. At the bottom of control tower the material is fed to a beater with the help of a pair of feed rollers. The beater treats the material against the grid bars for cleaning and transfers it to the feed lattice of the hopper feeder. A spiked lattice and an evener roller in the Hoper feeder per-form the operation of blending, opening, and cleaning. A stripper roller delivers the material to the feed lattice of the scutcher through a condensing box. The scutcher consists of a regulating feed unit, pin beater, lap-forming cage, calendering unit, and lap-winding unit. The incoming feeding material in the form of a bulky sheet is checked lengthwise and widthwise by a regulating feeding system. Thus a uniform amount of material is fed to the beater. The beater further opens and cleans the mate-rial and delivers it to the cage, which makes a lap sheet. The calender rollers compact this lap sheet while shell rollers wind it on a lap rod. Thus a compact roll of lap sheet is delivered by the scutcher, which is transferred manually on a trolley to the next machine called the card.

In the latest blow room lines, the material from fine openers/cleaners is transferred to the card directly by a fan through a chute feed system which is attached at the back of the card. Figure 3.4 shows the latest cotton blow room line in which scutcher is ex-cluded and fine cleaner delivery pipes are directly connected to the cards.

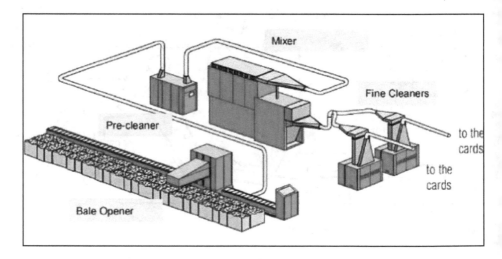

Fig. 3.4: Latest cotton blow room line.

3.3.3 Blow Room for Polyester

As polyester is man-made fiber so it has no trash particles like cotton and thus there is no need for cleaning while processing the polyester through the blow room. A typi-cal polyester blow room line consists of only two machines: a blending hopper and a pin-type opener as shown in Fig. 3.5. Pin-type opener delivers the material directly to the card through a fan and chute feed.

Objectives of a polyester blow room line are as follows:

– Opening: To open the compressed fibers up to very small tufts.
– Mixing and blending: To make a homogenous mixture of the material.
– Uniform feed for card: To provide output in the form of tufts of optimum size.

Fig. 3.5: Typical polyester blow room line.

3.3.4 Card

The conversion of tufty material into a single fiber state (filmy web) by passing it between wiry surfaces is called carding. Carding is a mechanical process that disentangles, cleans, and intermixes fibers to produce a continuous web or sliver suitable for subsequent processing. This is achieved by passing the fibers between differentially moving surfaces covered with card clothing. It breaks up locks and unorganized clumps of fiber and then aligns the individual fibers to be parallel with each other.

The objectives of carding process are as follows:

– Opening upto individual fibers
– Elimination of impurities
– Disentanglement of neps
– Elimination of dust
– Elimination of short fibers
– Fiber blending
– Fiber orientation or alignment
– Sliver formation

The following is the operation summary to achieve the tasks of card:

Lap prepared from the blow room is placed on the lap roller. The flow of material through the chute feed card is shown in Fig. 3.6. Feed roller with the help of a feed plate delivers it to the taker-in. The taker-in opens and cleans the cotton by dragging it on the mote knives and carding elements. Then this cleaned material, in the form of tiny tufts and single fibers, is transferred to the cylinder. Cylinder further cleans the cotton and converts it into the individual fiber state with the help of stationary and movable flats. At this stage, neps are disentangled and short fibers are separated from cotton. A fan sucks all the waste as well as dirt and dust from the whole machine and collects them in a box.

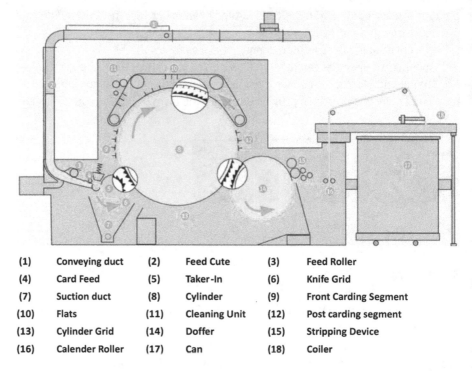

(1)	Conveying duct	(2)	Feed Cute	(3)	Feed Roller
(4)	Card Feed	(5)	Taker-In	(6)	Knife Grid
(7)	Suction duct	(8)	Cylinder	(9)	Front Carding Segment
(10)	Flats	(11)	Cleaning Unit	(12)	Post carding segment
(13)	Cylinder Grid	(14)	Doffer	(15)	Stripping Device
(16)	Calender Roller	(17)	Can	(18)	Coiler

Fig. 3.6: Material flow through chute feed card [3].

The doffer takes the fibers from the cylinder in fleece form. From the doffer, the fleece runs through the doffing roller, crush rollers, and tongue-groove rolls. These rollers scanned the cross-section/thickness of the carding sliver. The recorded results are compared with the set target value of sliver linear density. Deviations from the set value are corrected by altering the speed of the feed roller. From there, sliver in its final form is coiled into the cans with the help of coiler.

3.3.5 Difference Between Blow Room and Card Cleaning

In the blow room the tightly packed bales are converted into small tufts. These tufts are then transported step by step through a series of machines installed in a sequence. The coarse and gentle opening at the start of the blow room line is converted to the intensive and fine opening and cleaning at the end of the line. As the larger tuft are further converted into smaller tufts, the degree of cleaning increases gradually due to the generation of new surfaces and decreasing the density of the material. In blow room the opening and cleaning are performed at the same time by a combined action of air currents, opening spikes, and cleaning devices. The composition of trash, dust, fiber fragments, and fibers removed is called waste. In the case of a card machine, the material is opened to the individual fiber state. So, there are more chances of elimination of impurities, dust, and short fibers. In the taker-in zone, the coarse trash along with dust is separated from the material. While in carding zone the improved elimination of dirt and dust along with the removal of short fibers is carried out with the help of carding elements, mote knives, guiding elements, and suction tubes. The waste of cards is categorized into licker-in waste and fly waste.

3.3.6 Drawing

The following are the objectives of drawing:
- Equalizing: To improve the evenness of the sliver by doubling
- Parallelizing: To create a parallel arrangement of fibers in the sliver by drafting
- Blending: To compensate the raw material variations by doubling
- Dust removal: To remove dust within the overall process by suction
- Sliver formation: To make sliver and coil in a can by condensing and calendering

The following is the operation summary to achieve the objectives of drawing:

Four or eight sliver cans prepared on the card are arranged under the creel rollers of the drawing. The flow of material through the draw frame is shown in Fig. 3.7. Creel rollers withdraw the slivers from the cans and feed them to the drafting system. The slivers running into the drafting arrangement are attenuated by a draft of four to eight and a web having less cohesion is delivered. To avoid disintegration of the web, it is condensed into a sliver immediately after the drafting arrangement. This sliver is then guided through a pair of calender rollers and a tube that coil it into a can.

For cotton, normally two passages of draw frame are given while for blends of cotton with synthetic fibers three passages are used. Drawing frames may of single delivery or double delivery. Nowadays, single-delivery draw frames are used for final passage of material which is shown in Fig. 3.8.

3.3.7 Lap Former

The objectives of lap former are as follows:
- Equalizing: To improve the evenness of the lap by doubling of slivers.
- Parallelizing: To create a parallel arrangement of fibers in the lap by drafting of slivers.
- Blending: To compensate the raw material variations by doubling of slivers.
- Dust removal: To remove dust within the overall process by suction.
- Lap formation: To make lap by calendering.

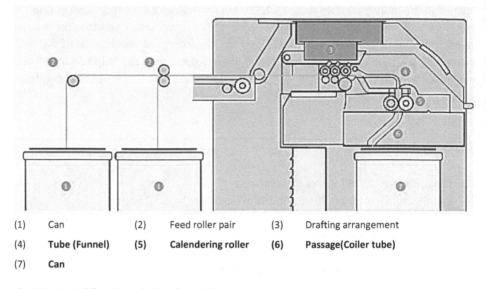

(1)	Can	(2)	Feed roller pair	(3)	Drafting arrangement
(4)	**Tube (Funnel)**	**(5)**	**Calendering roller**	**(6)**	**Passage(Coiler tube)**
(7)	**Can**				

Fig. 3.7: Material flow through draw frame [4].

Fig. 3.8: Draw frame.

The following is the operation summary to achieve the objectives of lap former:

Twenty-eight cans of drawn sliver from first draw frame are placed under the two creel rails of the lap former. The flow of material through the lap former is shown in Fig. 3.9. Creel rollers withdraw the slivers from the cans and feed them to the drafting system. The slivers running into the drafting arrangement are attenuated by a draft of 1.3–2.5.

The two webs created by the drafting system pass over two deflecting plates onto the web table. These webs are superimposed or placed one above the other. The calender

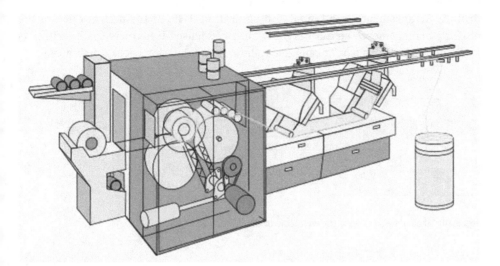

Fig. 3.9: Material flow through Unilap [4].

rollers draw these superimposed webs from the table and compact them to make a lap and deliver it to the lap-winding assembly. Winding assembly winds the lap on an empty tube. Empty tubes are automatically exchanged when the length of the lap is completed.

3.3.8 Comber

The objectives of comber are as follows:
- Noil removal: To remove short fibers, neps, and impurities by combing.
- Equalizing: To improve the evenness of the sliver by doubling.
- Parallelizing: To create a parallel arrangement of fibers in the sliver by drafting.
- Blending: To compensate the raw material variations by doubling.
- Dust removal: To remove dust within the overall process by suction.
- Sliver formation: To make sliver and coil in a can by condensing and calendering.

The following is the operation summary to achieve the objectives of comber:

Eight laps made on lap former are placed on the support rolls of the comber which unwind the laps very slowly and deliver to the feed roller. Assembly of nippers takes the lap from the feed roller. Circular combs comb the lap fringe, hanging from the nippers, and thus remove the short fibers, neps, and impurities. Nippers transfer the combed fibers to the detaching rollers which deliver it in the web pan.

This web is condensed into a sliver with the help of draw-off rollers, trumpet, and table calender rollers. Eight slivers coming from each lap are arranged parallel on the table and fed to the drafting system as shown in Fig. 3.10. The slivers running into the drafting arrangement are attenuated by a draft of 9–16. At the delivery end of the drafting arrangement, the discharged web is condensed into a sliver. This sliver is then guided through a pair of calender rollers and a tube which coils it into a can.

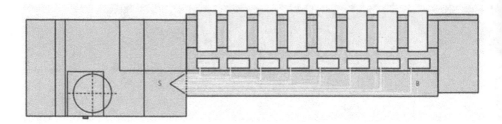

Fig. 3.10: Guiding the sliver from the web table to the drafting arrangement [4].

3.3.9 Roving Machine

The following are the objectives of a roving frame:
- Drafting to attenuate the sliver up to the required fineness.
- Twisting to impart strength
- Winding to make a roving package

The following is the operation summary to achieve the objectives of the roving frame:

Sliver cans from second drawframe or finisher draw frame are placed under the creel of the roving frame. The flow of material through the roving frame is shown in Fig. 3.11. Creel rollers withdraw the slivers from the cans and forward them to the drafting arrangement. The drafting arrangement attenuates the slivers with a draft of about 5–20. The material exiting from the drafting system is too thin and it cannot withstand itself. The twist inserting step is necessary immediately at the exit of the drafting arrangement in order to impart strength.

Twist insertion is done by the rotating flyer, usually in the range of 25–70 turns per meter. Roving from the flyer top runs through the hollow flyer leg and reaches the wind-up point with the help of the presser arm. Winding of roving is carried out due to the higher speed of the bobbin than the flyer. The roving coils are arranged on the bobbin very closely and parallel to one another by the bobbin rail which moves up and down continuously. The speed of the bobbin is reduced as the bobbin diameter is increased in order to keep its surface speed constant. Similarly, bobbin rail speed is decreased with the increase in bobbin diameter to maintain the coils per inch constant throughout the package building. During package building, the length of each next layer of roving is reduced continuously both from bottom and top in order to insert the taper on both ends. Figure 3.12 shows the roving frame which is being used in the industry.

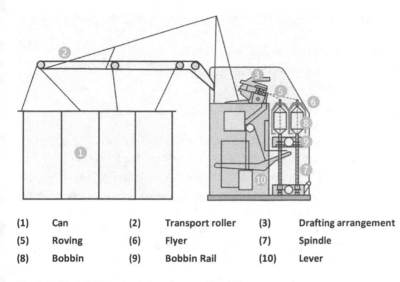

(1)	Can	(2)	Transport roller	(3)	Drafting arrangement
(5)	Roving	(6)	Flyer	(7)	Spindle
(8)	Bobbin	(9)	Bobbin Rail	(10)	Lever

Fig. 3.11: Material flow through roving machine [4].

Fig. 3.12: Roving machine.

3.4 Spinning Operations

There are a number of spinning techniques which are being used to produce spun yarns. However, three techniques (ring-spinning, open-end rotor spinning, and air-jet spinning) are common in the industry which will be discussed in the following subsections:

3.4.1 Ring Spinning

The following are the tasks which are required to achieve from ring spinning:
–   Drafting to attenuate the roving up to the required fineness
–   Twisting to impart strength
–   Yarn winding to make a suitable package

The following is the operation summary to achieve the tasks of the ring frame:

Roving packages made on roving frame are positioned on the hangers of the ring frame creel. Flow of material through the ring frame is shown in Fig. 3.13. Roving is fed to drafting system through guiding rods and roving guides. The drafting arrangement attenuates the roving with a draft required to make the final yarn count.

The ribbon exiting from the drafting system is too thin and it cannot withstand itself. So twist-inserting step is necessary to impart strength immediately. This step is

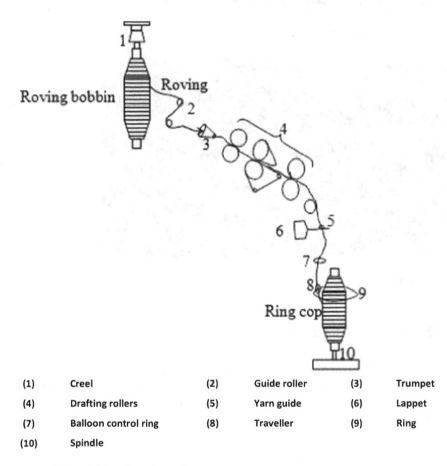

(1)	Creel	(2)	Guide roller	(3)	Trumpet
(4)	Drafting rollers	(5)	Yarn guide	(6)	Lappet
(7)	Balloon control ring	(8)	Traveller	(9)	Ring
(10)	Spindle				

Fig. 3.13: Material flow through ring frame.

performed by the ring and traveler with the help of the spindle. In this process, each rotation of the traveler on the spinning ring produces a twist in the yarn.

The ring traveler has no drive of its own; it is dragged with a spindle via the yarn attached to it. Winding of yarn on the bobbin is carried out due to the higher speed of the spindle than the traveler. The yarn is wound up into a cop form by raising and lowering of ring rail. Figure 3.14 shows the ring frame which is being used in the industry.

Fig. 3.14: Ring frame.

3.4.2 Winding

Winding is the creation of large yarn packages that can be easily unwounded. This makes easier and economical use of yarn on subsequent machines. Thus all yarns made on ringframe are wound in the form of large cones on autocone-winding machine. Yarn faults are also removed on this machine with the help of yarn clearer. Figure 3.15 shows the autowinding machine which is being used in the industry.

3.4.3 Open-End Rotor Spinning

In open-end rotor-spinning process the preparatory processes include the operations of blow room, card, and draw frame passage. For the spinning of coarser yarn counts with shorter fiber lengths, card sliver can directly be fed to the rotor machine. However the need for the right quality of sliver determines the requirement of one or two draw frame passages after carding process. Instead of the classical roller-drafting technique, dispersion drafting is used in rotor machines. The twist is also inserted due to the rotation of rotor.

Fig. 3.15: Winding machine.

Usually, the sliver cans from the first draw frame are placed under the open-end rotor machine. The flow of material through open-end rotor machine is shown in Fig. 3.16. The sliver from the can is fed to the feed roller with the help of a sliver guide. Combing roller takes the sliver from the feed roller and opens it up to individual fiber and delivers these fibers to the rotor through a fiber transfer tube. The fibers are deposited onto the rotating rotor and slide down into the rotor groove and form a ribbon of fibers. The rotor rotates at a very high speed creating a centrifugal force due to which the rotor is under a partial vacuum.

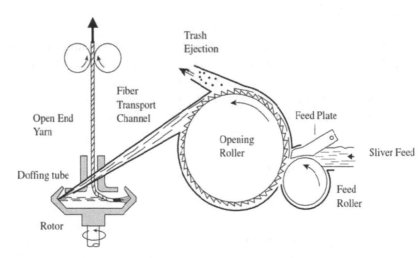

Fig. 3.16: Material flow through open-end rotor machine.

To start spinning, a length of yarn already wound onto the package of the take-up mechanism is threaded through the nip line of the delivery rollers and into the draw-off tube. Because of the partial vacuum, the tail end of this yarn is sucked into the rotor due to vacuum. The rotation of the rotor pulls the yarn end onto a collected ribbon of fibers and simultaneously inserts the twist into the yarn tail. A little of this twist propagates into that part of the ribbon in contact with the yarn tail, binding it to the yarn end. Once the yarn tail enters the rotor, the delivery roller is set in motion to pull the tail out of the rotor. The pulling action on the tail results in peeling of the fiber ribbon from the rotor groove. The newly formed yarn is wound up on the package by a winding drum.

The production rate of rotor spinning is six to eight times higher than that of ring spinning. Open-end rotor machines are fed directly by sliver and yarn is wound onto packages ready for use in fabric formation. So in open-end rotor spinning, only one machine is used instead of three machines (roving frame, ring frame, and auto winding) in ring spinning. Rotor-spun yarns are more even but somewhat weaker and have a harsher feel than ring-spun yarns.

Rotor-spun yarns are mainly produced in the medium count (40 Ne, 20 tex) to coarse count (05 Ne, 60 tex) range. End uses include denim, towels, blankets socks, T-shirts, shirts, and pants. Figure 3.17 shows the latest open-end rotor machine which is being used nowadays in the industry.

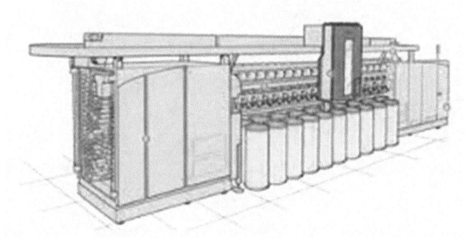

Fig. 3.17: Rotor-spinning machine [5].

3.4.4 Air-Jet Spinning

Flow of material through the air-jet spinning machine is shown in Fig. 3.18. In order to have adequate parallelization of fibers required for air-jet spinning the sliver, which has passed through three passages of draw frame, is preferred as infeed material. The slivers coming from the creel portion are passed over the stationary creel rods being directed to the drafting arrangement. The drafting arrangement permits a draft of 100–200 depending upon the required yarn fineness. The highly attenuated fiber strand then passed through the spinning nozzles. The false twist is imparted by the power air vortex generated by the nozzles. The direction of the air vortex in these two nozzles is opposite to each other. A typical nozzle arrangement is shown in Fig. 3.19.

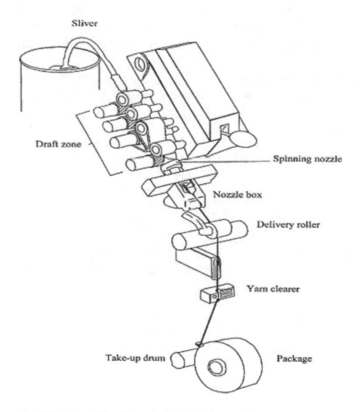

Fig. 3.18: Material flow through air-jet spinning machine.

Due to the low intensity of the first jet, it only affects the small number of edge fibers. Thus it wraps the edge fibers around the core fibers. The angular velocity of the air vortex inside the second jet is more than 1 million rpm which inserts the twist to all the edge fibers and wound around the parallel fiber strand. They bind the body of fibers together and ensure coherence. The resultant yarn is cleared of any defects and wound onto packages.

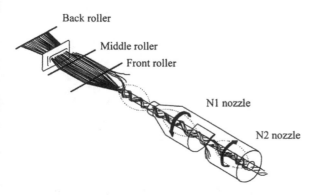

Fig. 3.19: Two-nozzle arrangement.

The production rate of air-jet/vortex spinning is three to five times higher than rotor-spinning and 10–20 times that of ring-spinning machine. Just like rotor-spinning, the air-jet spun yarn is very cheaper to produce since it also uses fewer production stages. As is the case with rotor-spun yarns, the air-jet spun yarns are more even, but weaker and have a harsher feel than that of ring-spun yarns. The air-jet spun yarns are mostly produced in the medium count (30 Ne, 20 Tex) range and are mainly polyester/cotton blended yarns. End users of vortex-spun yarn include woven sheets and knitted light-weight shirting. Figure 3.20 shows the latest air-jet machine which is being used nowadays in the industry.

Fig. 3.20: Air-jet spinning machine.

3.5 Types of Yarns

There are three types of yarns:
- Staple spun yarns
- Monofilament yarns
- Multifilament yarns

3.5.1 Staple Spun Yarn

- Is made from staple fibers – cotton, wool, or manufactured fibers cut into short lengths
- Is uneven, weak yarn with poor luster and durability
- Have good elasticity, resiliency, and absorbency
- Are used mainly for apparel and furnishings

3.5.2 Monofilament Yarn

- Monofilaments are simply single filaments of synthetic fibers that are strong enough to be useful without being twisted with other filaments into a yarn.
- They are fine and strong with good luster and durability but are inelastic in nature with poor resiliency and absorbency.
- Monofilament yarns are used primarily for hosiery and invisible sewing thread.

3.5.3 Multifilament Yarn

- Is made from two or more filaments of a manufactured fiber.
- Is an even, strong yarn with good luster and durability.
- Has medium elasticity and resiliency and is slightly absorbent.
- Is used primarily for evening wear and lingerie.

3.6 Yarn Quality Parameters

To control the quality of yarn in a spinning mill, quality checks are carried out at every stage of manufacturing starting from raw cotton to each step of yarn manufacturing to meet the customer standards. A plan of production requires certain standard levels to which materials in the process must conform. Quality may be defined as "the degree of excellence of any product or it is to meet the customer expectations." In a spinning mill, the following parameters are maintained with respect to products at each stage of processes:

Blow room: Lap weight per yard and its variation, lap rejection, fiber growth, extracted waste percentage, and cleaning efficiency

Card: Sliver weight per yard and its variation, sliver mass variation, nep count of the web, extracted waste percentage, fiber growth, and cleaning efficiency

Drawing: Sliver weight per yard and its variation, sliver mass variation, and fiber growth

Lap former: Lap weight per yard and its variation

Comber: Sliver weight per yard and its variation, sliver mass variation, extracted noil percentage and its analysis, and fiber growth

Simplex: Roving count and its variation, roving mass variation, and fiber growth

Ring: Yarn count and its variation, lea strength and its variation, CLSP, single yarn strength, elongation, unevenness, thin places, thick places, neps, hairiness, and twist

Autocone: Yarn count and its variation, lea strength and its variation, CLSP, single yarn strength, elongation, unevenness, thin places, thick places, neps, hairiness, and twist

References

[1] Celanese Acetate. *Complete Textile Glossary*. Three park avenue New York, USA: Celanese Acetate LLC, 2001.
[2] M. J. Denton, and P. N. Daniels, *Textile Terms and Definitions*. UK: The Textile Institute, 11th ed. 2002.
[3] W. Klein, *The Rieter Manual of Spinning Volume 2 – Blow Room & Carding*. Klosterstrasse 20, CH-8406, Wintherthur: Rieter Machine Works Ltd AG, 2014.
[4] W. Klein, *The Rieter Manual of Spinning Volume 3 – Spinning Preparation*. Klosterstrasse 20, CH-8406, Wintherthur: Rieter Machine Works Ltd AG, 2014.
[5] H. Ernst, *The Rieter Manual of Spinning Volume 5 – Rotor Spinning*. Klosterstrasse 20, CH-8406, Wintherthur: Rieter Machine Works Ltd AG, 2014.

Khubab Shaker, Muhammad Imran Khan

4 Woven Fabrics

Abstract: The fabrics are used for clothing as well as for industrial and technical applications. The end application depends on the fabric structure and the raw material used. The main categories of fabrics are woven, knitted, and nonwoven fabrics. All these types vary considerably in the process, production rate, and fabric structure. The woven fabrics are produced by the interlacement of two sets of yarns, namely warp and weft. Prior to interlacement, the warp yarn is prepared by a series of steps including winding, warping, sizing, and drawing in/knotting. The woven fabric is then produced by interlacing warp and weft on the loom, following a series of loom motions in a specific sequence. The fabric produced is inspected for any defects before dispatch.

Keywords: warp, sizing, loom motions, weft, inspection

4.1 Textile Fabrics

Textile fabric may be defined as a flexible assembly of fibers or yarns, either natural or man-made. A variety of techniques are used to produce textile fabric, but weaving, knitting, and nonwovens are the most common. Conventional fabrics (woven, knitted) are produced in such a way that the fibers are first converted into yarn and subsequently this yarn is converted into fabric. The fabrics can also be produced directly from the fibers. Such fabrics are termed nonwovens.

Weaving is the most used technique of fabric manufacturing. The woven fabrics find application in numerous areas like apparel, home textile, filters, geotextiles, composites, medical, packaging, seatbelts, industrial products, and protection. The history of weaving dates back to ancient times when human beings used woven fabrics to cover themselves. Several evidences indicate that Egyptians made woven fabrics around 6,000 years ago and that silk became an important economic commodity in China some 4,000 years ago [1].

The woven fabrics are produced by the interlacement of two sets of yarns perpendicular to each other [2], that is, warp and weft as shown in Fig. 4.1. The first set includes the threads running lengthwise in the fabric, while the second is represented by the threads placed in cross or width direction. The fabrics have varying structures, depending on the interlacement pattern of the yarns. This sequence of interlacements is called the weave design of the fabric. The properties of the fabric depend on various factors such as the inherent properties of fibers, a linear density of yarn, weave design, and the porosity of the structure.

https://doi.org/10.1515/9783110799415-004

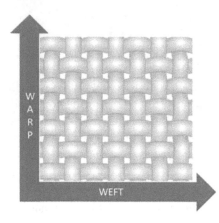

Fig. 4.1: Schematic of warp and weft in woven fabric.

4.2 Warp Preparation Steps

A summary of the process steps from yarn to the final product, that is, the loom-state fabric is shown in Fig. 4.2. The warp yarn is subjected to a number of processes, known as warp preparation before conversion into the fabric, while weft yarn does not require any specific preparation. The warp preparatory process consists of the following operations: winding, warping, sizing, and drawing-in or knotting.

Yarns produced in spinning are used as input for the warp preparation. Winding process helps to prepare the yarn packages according to the required shape and size. The weft yarn is then provided to loom, while warp yarns are processed to give a sheet of yarns on warp beam by the process called warping. The objective of warping is to reduce the yarn faults and get required number of threads in fabric. The next process is sizing in which a coating of size material is applied to the yarn to impart strength and make it smooth. All the sheets from warper beams are combined on the weaver's beam after sizing.

There are two possible ways (drawing in and knotting) once the sizing is done. As long as the weave design does not change, knotting is used. The old beam yarns are knotted with the new beam yarns. If the quality changes, the warp sheet is drawn in

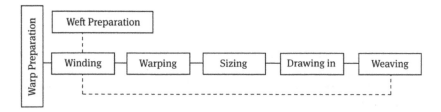

Fig. 4.2: Flow of the weaving process.

from the droppers, heald frames, and reeds, a process known as drawing in. The actual fabric forming process is carried out at the loom, where this warp sheet and weft are interlaced to produce the woven fabric.

4.2.1 Winding

Winding is a process in which yarn from bobbins, which is the end product of ring spinning, are converted into a suitable form of package. This transfer of yarn from one type of package to another package, more suitable for the subsequent process, is also called winding. The main objectives of winding process are to increase the package size, clear yarn defects, and produce a package suitable for subsequent process (size and shape).

The yarn packages are either parallel or tapered, with respect to shape, as shown in Fig. 4.3. The parallel packages may also have flanges, while tapered packages are without flange [1].

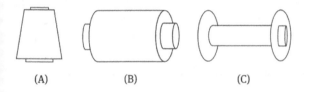

(A)   (B)   (C)

Fig. 4.3: Package types: (a) tapered, (b) parallel without flange, and (c) flanged parallel.

The winding process involved unwinding yarn from one package and rewinding it on to another package. The yarn may be unwound in two ways, that is, over end and side withdrawal as shown in Fig. 4.4. Winding rate is the speed at which the yarn is wound on package surface, while to and fro movement of yarn when it is laid on to package is called traverse. In case of near-parallel package, traverse is very slow, but in case of cross-wound package traverse is quick. There is no traverse in case of parallel wound packages.

In winding machine, yarn is taken from the bobbin/cop and is wound on the package after passing through the thread guides, balloon breaker, stop motion, and yarn clearer. For cross-winding, a grooved drum is also provided on the machine to traverse the yarn.

4.2.2 Warping

In warping process, the yarns are transferred from several supply packages (cones) to the warp beam in the form of a parallel sheet. The main objective of warping is to get the required number of ends as per requirement and remove/minimize the spinning

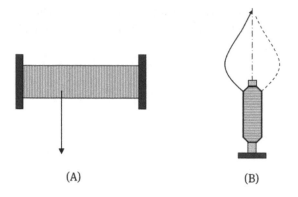

(A)   (B)

Fig. 4.4: Package unwinding: (a) side withdrawal and (b) over-end withdrawal.

faults. The three main types of warping are high speed/direct warping, sectional/indirect warping, and ball warping. In direct warping, the yarns are withdrawn from the single-end yarn packages (cone) placed on the creel and directly wound on a beam. Several beams are warped to get the required number of ends. For example, to produce a fabric with 6,040 warp ends, eight beams will be warped, each with 755 ends. These beams are then combined into a single beam in the sizing process. The direct warping process offers only limited pattern possibilities and is preferred for simple patterns only.

The indirect/sectional warping process completes in two steps called warping and beaming. In warping step, a portion of the required number of threads (called section) is wound onto a drum (Fig. 4.5). The subsequent sections are warped on the drum side by side, one after the other. In the next step all the sections are unwound from the drum and wound onto beam to complete the required number of threads. This beam may or may not be taken for the sizing process. The division of warp sheet into small sections provides unlimited patterning possibilities. Therefore, this process is suitable for complex warp patterns. Ball warping is the process in which warping is performed in rope form onto wooden ball. The ball is wound on a special wooden core called "log." It is also a two-stage process, suitable for denim fabric manufacturing, involving rope-dyeing process. Rebeaming is done to convert the rope-dyed warp yarn, stored in cans, into beam.

4.2.3 Sizing

Sizing also known as slashing is the coating of a warp sheet with a size solution. Weaving requires the warp yarn to be strong, smooth, and elastic to a certain degree. There is always friction between metallic parts and yarn during the weaving. So, the warp yarns need to be lubricated to reduce the abrasion. The application of size

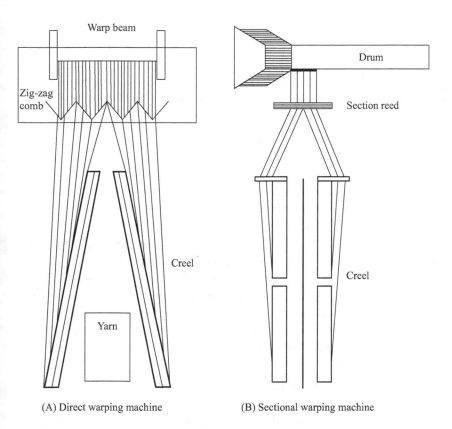

(A) Direct warping machine   (B) Sectional warping machine

Fig. 4.5: Direct and sectional warping machine schematic.

material helps to improve the mechanical properties of warp and reduce abrasion and the elasticity of the yarn. The amount of sizing material relates to the tenacity, hairiness, and linear density of yarn and also to its behavior during weaving. Another major objective of this process is to get the total ends on a weaver's beam, combining the ends of all warp beams. The application of sizing material results in the following properties in yarn:

– High strength
– Low flexibility
– Low abrasion
– Increased smoothness
– Less hairiness

The process of sizing can be classified based on the method of application into conventional wet-sizing, solvent-sizing, cold-sizing, and hot melt-sizing [3]. The main parts of a conventional sizing machine include (Fig. 4.6) creel, sizing box, drying section, leasing section, headstock, and size cooker.

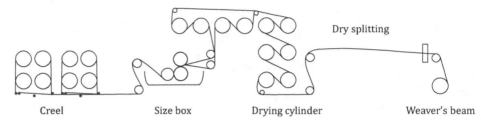

Fig. 4.6: Schematic diagram of conventional wet-sizing machine.

In conventional wet sizing, the fundamental constituents of the size recipe are the size materials and a solvent usually water. The sizing materials are broadly classified into three groups namely adhesives, softeners, and auxiliaries [4].

The adhesives perform two functions: bind the constituent fibers of the yarn together and form a film over the yarn surface, resulting in increased strength, low hairiness, and more even yarn. Depending on their origin, adhesives are classified as natural, synthetic, or modified adhesives. In order to make modified adhesives, natural adhesives are treated with certain chemicals. The natural adhesives may be obtained from plants or animals, for example, maize starch and potato starch. The chemical modification of natural adhesives is performed to induce the desired properties. Some common examples of modified adhesives are modified starches and carboxymethyl cellulose. The chemically synthesized polymers like polyvinyl alcohol and acrylics fall under the category of synthetic adhesives. Starch adhesives are used most commonly because of low cost and environment safety.

The softeners are added in the size recipe to lubricate the yarn and reduce abrasion/friction between adjacent yarns and between yarns and loom accessories. They also give a soft handle to the warp and size film, helping to decrease its brittleness. The softeners may be in solid form (wax group) or liquid form (oil group) and are obtained from animals, vegetables, or synthesized chemically. The auxiliaries include antiseptic, antistatic, weighting, swelling agents, and/or defoamers. The sized fabric must be subjected to a desizing process prior to the finishing stage. Desizing has a decisive effect on the wastewater load in textile production.

4.2.4 Drawing In

The sized warp sheet is wound on to a beam called as the weaver's beam. It has the required number of ends, and the yarns have adequate strength to bear the tensions of weaving process on loom. This beam is either used for drawing in or knotting/tying, depending on the requirement (Fig. 4.7).

Style change involves the production of a new fabric style, while mass production means to continue the weaving of same fabric style just replacing the empty beam

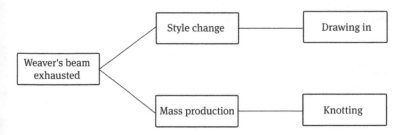

Fig. 4.7: Drawing in/knotting procedure.

with a full beam of same type. Drawing in is the process of entering the individual yarn of warp sheet through dropper, heald eye, and the reed dent (Fig. 4.8). The yarns can be threaded either manually or by using automatic machines.

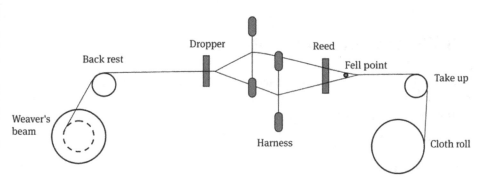

Fig. 4.8: Drawing in schematic and yarn path.

The yarn is now fully prepared for conversion into the fabric, which takes place at loom. The weaver's beam is gaited on the loom, while weft yarn is provided at right angle either from cone or bobbin depending on the picking media.

4.2.5 Fabric Inspection and Grading

Once the fabric has been produced on loom, it is taken to the folding department where it is inspected for quality and dispatched. The fabric is visually inspected on frames and each fault is given points according to the nature of the fault. The fabric faults are categorized into mendable and nonmendable faults. The mendable faults are repaired, while nonmendable faults are given the points and/or cut and separated from the fabric. The fault points are awarded following four-point system, as given in Tab. 4.1.

One single meter will not be given more than four points even if there are more than one faults of above 9". Some common faults include knots, slub, broken ends,

Tab. 4.1: Four-point grading system for woven fabrics.

Fault size	Points given
1″–3″	1
3″–6″	2
6″–9″	3
Above 9″	4
Hole up to 1″	2
Hole over 1″	4

double end, hole, cut, local distortion, missing ends, missing pick, oil and other stains, reed mark, snarls, and weft bar.

4.3 Weaving Mechanisms

The conversion of warp sheet into fabric by interlacing with weft yarn requires the basic operations to be carried out on loom in a specific order. It involves the primary motions, secondary motions, and the stop motions [5].

4.3.1 Primary Motions

The primary loom motions include the following three operations:
Shedding: The separation of the warp sheet into two layers to form a tunnel known as the shed

Picking: Insertion of weft yarn, across the warp sheet width, through the shed

Beat-up: Pushing the newly inserted length of weft (pick) to the fell of cloth

These operations occur in a given sequence and their precise timing in relation to one another is of extreme importance.

4.3.2 Secondary Motions

The secondary motions facilitate the weaving of fabric in a continuous way [6]. These include:

Let off: This motion provides warp sheet to the weaving area at the required rate and under constant tension by unwinding it from weaver's beam

Take-up: This motion draws fabric from the weaving area at a uniform rate to produce the required pick spacing and wind it onto a roller

4.3.3 Stop Motions

These motions are used in the interest of quality and productivity, stopping the loom immediately in case of some problem. The warp-stop motion will stop the loom in case any warp-yarn breaks, avoiding excessive damage to the warp threads. Similarly, weft-stop motion will come into action in the absence of weft yarn and stop the loom.

4.4 Types of Shedding Mechanism

There are three most common types of shedding mechanisms, namely tappet, dobby, and jacquard shedding [7]. Tappet and dobby systems control heald frames while jacquard provides control of individual warp yarn.

4.4.1 Tappet Shedding

This system is also termed as cam shedding. The cam is an eccentric disc mounted on the bottom shaft, rotating to lower or lift the heald frame. It is relatively simple and inexpensive system handling up to 14 heald frames [8]. But this system has very limited design possibilities and pick repeat, producing simple weaves.

4.4.2 Dobby Shedding

It is a relatively complex shedding system and can control up to 30 heald frames. The pick repeat to dobby system is provided by peg chain, punched papers, plastic pattern cards, or computer programming and is virtually unlimited. This system offers more design possibilities as compared to tappet shedding.

4.4.3 Jacquard Shedding

The jacquard shedding provides unlimited patterning possibilities. The working principle is relatively simple but involves a greater number of parts that make it a complex machine. Versatility of jacquard shedding is due to control over individual warp yarn. The jacquard shedding system can be either mechanical or electronic.

4.5 Types of Picking Mechanism

Picking involves the insertion of the weft yarn through shed across the width of warp sheet. The picking mechanism is mainly a function of the picking media used for the insertion of weft (Fig. 4.9). Weft velocity and insertion rate determine the picking media, which are classified into shuttle and shuttle-less picking.

4.5.1 Shuttle Picking

It is the oldest technique of weft insertion on loom. The picking media is a wooden shuttle that traverses back and forth across the loom width. A pirn or quill having yarn wound on it is placed inside the shuttle. As the shuttle moves across, the yarn is unwound and placed in the shed. A picking stick on each side of loom helps to accelerate the shuttle by striking it. Shuttle travels on the race board above lower portion of the warp sheet. The shuttle picking takes place from both the sides of loom.

4.5.2 Projectile Picking

Introduced first time by Sulzer in 1952, this machine uses a small metallic projectile along with gripper to throw the weft yarn across the loom width. The energy required for propulsion of projectile into the shed is provided by twist in the torsion rod. The projectile glides through guide teeth in the shed. It had low power consumption, versatility of yarns, and a higher weft insertion rate as compared to the shuttle picking system.

4.5.3 Rapier Picking

This picking system uses a rigid or flexible element called rapier for the insertion of weft yarn across the shed. There are two major variations in the rapier picking: single rapier and double rapier. In case of single-rapier picking system, the rapier head grips the weft and carries it across the shed to receiving end. The rapier has to return empty to insert the new weft. The double-rapier picking makes use of two rapiers [9]. One rapier (giver) takes yarn to the center of machine and transfers it to the other rapier (taker), which brings the weft to the other side.

4.5.4 Water Jet Picking

The water jet picking involves the insertion of weft yarn by highly pressurized water. This pressurized water takes the form of a coherent jet due to the surface tension viscosity of water. The flow of water has three phases: acceleration inside pump, jet outlet from nozzle, and flow into the shed. The amount of water used for the insertion of one pick is less than 2 cm^3. This system is mostly preferred for the synthetic yarns.

4.5.5 Air Jet Picking

In air jet picking system, the weft is inserted into the shed using compressed air. The yarn is taken from the supply package/cone and wound on to the feeder before insertion to avoid tension variations. The weft is then passed through the main nozzle which provides initial acceleration to the yarn. The auxiliary nozzles are present at specific distance along the width to assist in weft insertion. A special type of reed, called profiled reed, is used for air jet picking. The channel in the reed guides the yarn across the shed and avoids entanglement with warp. It has an extremely high weft insertion rate.

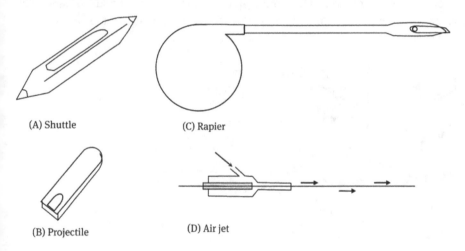

(A) Shuttle

(C) Rapier

(B) Projectile

(D) Air jet

Fig. 4.9: Mechanisms of weft insertion.

4.6 Weave Design

The woven fabric is produced by interlacement of warp and weft, and this interlacement pattern is called weave design of the fabric [10]. The three basic weave designs are plain, twill, and satin.

4.6.1 Plain

The simplest interlacing pattern for warp and weft threads is over one and under one as shown in Fig. 4.10. The weave design resulting from this interlacement pattern is termed as plain or 1/1 weave. The 1/1 interlacement of yarns develops more crimp and fabric produced has a tighter structure. The plain weave is produced using only two heald frames. The variations of plain weave include warp rib, weft rib, and matt or basket weave.

4.6.2 Twill

This weave is characterized by diagonal ribs (twill line) across the fabric. It is produced in a stepwise progression of the warp yarn interlacing pattern. The interlacement pattern of each warp starts on the next filling yarn progressively. The two subcategories based on the orientation of twill line are Z- and S-twill or right-hand and left-hand twill, respectively. Some of the variations of twill weave include pointed, skip, and herringbone twill [11].

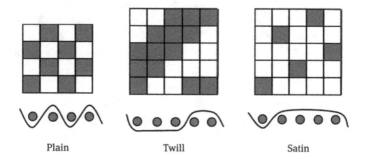

Plain   Twill   Satin

Fig. 4.10: Basic weaves and their cross-sectional view.

4.6.3 Satin/Sateen

The satin/sateen weave is characterized by longer floats of one yarn over several others. The satin weave is warp-faced while sateen is a weft-faced weave. A move number is used to determine the layout in a weave repeat of satin, and a number of interlacements are kept to a minimum as shown in Fig. 4.10. The fabrics produced in satin weave are more lustrous as compared to corresponding weaves.

4.7 Specialty Weaving

There are certain specialty weaving techniques used for the production of a specific fabric type, for example, circular loom, terry towel loom, denim fabric, narrow loom, multiphase loom, three-dimensional (3D) weaving loom, and carpet/rug weaving. The weaving is also used for the production of certain industrial fabrics and technical textiles [2] like conveyor belt fabrics, air bag fabrics, cord fabrics, geotextiles, ballistic protection, and tarpaulins. The denim fabrics are woven with a coarse count, high thread density, and 3/1 twill weave. Dyeing is an additional process involved in the warp preparation for these fabrics [12]. The warp yarn of these fabrics is dyed with indigo dyes in such a way that only surface is dyed and core remains white. The narrow loom usually involves a needle for the weft insertion. It usually draws the warp sheet directly from the creel through tensioning rollers, thus helping to increase efficiency and productivity.

The towels are piled fabrics produced from two different set of warps: one serving as the ground and the other as pile. More length of pile warp is consumed as compared to ground warp. Therefore, two beams are required to produce such fabrics and need additional attachments on loom. In multiphase loom, several weft yarns are inserted simultaneously across the series of sheds. These sheds are arranged sequentially in the warp direction. The 3D loom produces a 3D fabric on the required shape [13]. The carpet weaving involves a loom with two beam arrangement as in case on terry towel fabrics. The ground warp let-off, pile warp let-off, and cloth take-up are controlled by servomotors [14]. It allows easy change of pile height and pick density. The tension in the pile warp sheet is controlled by a pneumatic beam brake.

References

[1]   P. R. Lord, and M. H. Mohamed, *Weaving: Conversion of Yarn to Fabric*. Durham, England: Merrow Publishing 1982.

[2]   S. Adanur, *Handbook of Weaving*. Washington, USA: CRC Press, 2001.

[3]   B. Wulfhorst, O. Maetschke, M. Osterloh, A. Busgen, and K.-P. Weber, *Textile Technology*. Weinheim: Wiley Online Library, 2006.

[4]   B. C. Goswami, R. D. Anandjiwala, and D. Hall, *Textile Sizing*. New York: CRC Press, 2004.

[5]   R. Marks, and A. T. C. Robinson, *Principles of Weaving*. Manchester: The Textile Institute, 1976.

[6]   A. Talukdar, *Sriramulu, Weaving Machines Mechanism Management*. 2nd ed. Ahmedabad: Mahajan Publishers, 1998.

[7]   J. Arora, *Modern Weaving Technology*. India: Abhishek Publications, 2002.

[8]   A. Ormerod, *Modern Preparation and Weaving Machinery*. England: Woodhead Publishing, 2004.

[9]   S. Maity, K. Singha, and M. Singha, "Recent developments in rapier weaving machines in textiles," American Journal of System Science, vol. 1, no. 1, pp. 7–16, 2012.

[10] R. Marks, and A. T. C. Robinson, *Woven Cloth Construction*. Manchester: The Textile Institute, 1973.

[11] E. B. Berry, *Textile Designing: Pure and Applied*. North Carolina, USA: North Carolina State College, 1967.

[12] S. Adanur, "Property analysis of denim fabrics made on air-jet weaving machical part II: Effects of tension on fabric properties," Textile Research Journal, vol. 78, no. 1, pp. 10–20, Jan 2008.

[13] N. Gokarneshan, and R. Alagirusamy, "Weaving of 3D fabrics: A critical appreciation of the developments," Textile Progress, vol. 41, no. 1, pp. 1–58, Apr 2009.

[14] A. M. Seyam, Advances in weaving and weaving preparation, Textile Progress, vol. 30, no. 1–2, pp. 22–40, 2000.

Muhammad Umair

5 Nonwovens

Abstract: Nonwoven fabrics are used for clothing as well for hygienic, industrial, and different technical applications. The area of end use depends on the raw material and method of manufacturing. During conventional weaving and knitting fabric formation methods, different processes are required for yarn preparation. However, in nonwoven fabrics, fibers are directly converted into the fabric. Hence their production rate is very high in comparison to the conventional fabric formation techniques. Along with two-dimensional fabrics, three-dimensional nonwoven fabric assemblies, having substantial thickness, are also possible. The properties of nonwoven fabrics are governed by their formation technique, fiber type, fiber content, and type of finish applied.

Keywords: nonwoven, **INDA**, EDANA, web formation, web bonding, finishing

5.1 Introduction

The textile fabric may be defined as a flexible assembly of fibers or yarns, either natural or man-made. It may be produced by several techniques, the most common of which are weaving, knitting, and nonwovens. Conventional fabrics (woven and knitted) are produced in such a way that the fibers are first converted into yarn and subsequently this yarn is converted into fabric. The fabrics can also be produced directly from the fibers. Such fabrics are termed as nonwovens. Nonwovens can be produced by different methods and each of these methods is capable of producing a large number of fabric structures, depending upon the raw material, machinery, and the process involved. These fabrics are used for a wide range of applications from clothing to technical purposes.

During nonwoven fabric production, the yarn manufacturing as well as yarn preparation processes (required in woven fabric) are eliminated. Due to this reason nonwoven fabrics are cheaper as compared to conventional fabrics. The great advantage of nonwoven fabrics is the speed with which the final fabric is produced. All yarn preparation steps are eliminated, and the fabric production itself is faster than the conventional methods. Not only the production rate is higher for nonwovens as shown in Tab. 5.1 [1] but also the process is more automated, requiring less labor than even most modern knitting or weaving systems. The nonwoven process is also energy-efficient.

In the nineteenth century, it was realized that a large amount of fiber is wasted as trim. A textile engineer named Garnett developed a special carding device to shred this waste material back into fibrous form. This fiber was used as filling material for

https://doi.org/10.1515/9783110799415-005

Tab. 5.1: Production comparison of woven, knitted, and nonwoven fabrics (INDA).

Method	System	Production (m^2/h)
Weaving	Shuttle	15
Overall average 0.583 m^2/min	Rapier	30
	Water jet	35
	Projectile	40
	Air jet	55
Knitting	Double knit	125
Overall average 8.5 m^2/min	Rib	175
	Single jersey	250
	Raschel	800
	Tricot	1,200
Nonwoven	Stitch bonded	450
Overall average 335.5 m^2/min	Needle punched	7,200
	Card bond	15,000
	Wet laid	30,000
	Spun laid	48,000

pillows. The Garnett machine though greatly modified, today still retains his name and is a major component in the nonwoven industry. Later on, manufacturers in Northern England began binding these fibers mechanically (using needles) and chemically (using glue) into batts, which are used as the precursors of today's nonwovens [2].

The term "nonwoven" rose more than 60 years ago when nonwovens were considered a cheaper alternative to conventional textiles and were generally made from carded webs using textile-processing machinery [2]. Nonwoven industry is very sophisticated and profitable, with healthy annual growth rates. It is perhaps one of the most intensive industries in terms of its investment in new technology and also in research and development. Therefore, the nonwoven industry, as we know it today, has grown from developments in the textile, paper, and polymer processing industries. Today, there are also inputs from other industries including most branches of engineering as well as the natural sciences.

5.1.1 Definitions

Different definitions of nonwovens are available by different organizations. According to the ASTM D 1117–01, nonwovens can be defined as follows:

> A textile structure produced by the bonding or interlocking of fibers, or both, accomplished by mechanical, chemical, thermal or solvent means and combinations thereof.

According to the standard ISO-9092:1988, the nonwovens are:

Manufactured sheet, web or batt of directionally or randomly orientated fibers, bonded by friction, and/or cohesion and/or adhesion, excluding paper and products which are woven, knitted, tufted, stitch-bonded incorporating binding yarns or filaments, or felted by wet-milling, whether or not additionally needled. The fibers may be of natural or man-made origin.

The Association of Nonwoven Fabrics Industry, USA (INDA) defines nonwovens as follows:

A sheet, web or batt of natural and/or man-made fibers or filaments, excluding paper, that have not been converted into yarn, and that are bonded to each other by any of several means.
    To distinguish wet-laid nonwovens from wet-laid paper materials the following differentiation is made. (a) More than 50% by mass of its fibrous content is madeup of fibers with a length to diameter ratio greater than 300. Other types of fabrics can be classified as nonwoven if, (b) More than 30% by mass of its fibrous content is made up of fibers with a length to diameter ratio greater than 600 and/or the density of the fabric is less than 0.4 g/cm³.

The European Disposables and Nonwovens Association (EDANA) describes the nonwovens as follows:

A manufactured sheet, web or batt of directionally or randomly oriented fibers, bonded by friction, and/or cohesion and/or adhesion, excluding paper and products, which are woven, knitted, tufted or stitch-bonded, or felted by wet-milling, whether or not additionally needled. The fibers may be of natural or man-made origin. They may be staple or continuous filaments or be formed in situ". To distinguish wet-laid nonwovens from wet-laid papers, a material shall be regarded as a nonwoven if, (a) More than 50% by mass of its fibrous content is made-up of fibers with a length to diameter ratio greater than 300; or (b) More than 30% by mass of its fibrous content is made up of fibers with a length to diameter ratio greater than 300 and its density is less than 0.40 g/cm³.

5.1.2 Nonwoven Products

Nonwoven fabrics have a wide area of applications depending upon the properties required in the end product. EDANA has given the nonwoven fabric applications according to the end use as shown in Fig. 5.1.

Nonwoven fabrics have a wide range of products including disposable gowns, face masks, gloves, shoe covers, dressings, sponges, wipes, diapers, sanitary napkins, lens tissues, vacuum cleaner bags, tea and coffee bags, hand warmers, interlinings, incontinence products, floor covers, air filters, paddings, blankets, pillows, pillowcases, aprons, table cloths, handbags, book covers, posters, banners, disk liner, sleeping bags, tarpaulins, tents, crop covers, greenhouse shadings, weed control fabrics, golf and tennis courts, road beds, drainage, sedimentation and erosion control, soil stabilization, and dam embankments [4, 5].

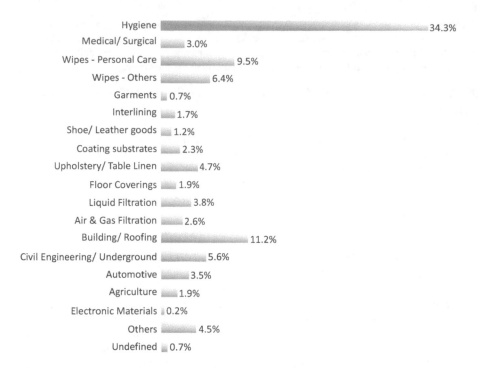

Fig. 5.1: Application areas of nonwovens [3].

5.1.3 Raw Materials for Nonwovens

Different types of man-made and natural fibers are used in nonwoven fabric production. Man-made fibers have 90% share of total fiber consumption [2].

Polymers, fibers, and binders are the basic raw materials for nonwovens. Most of the fibers and binders are made of polymers. A polymer is a large molecule built up by the repetition of small single chemical units. Large molecules are called macromolecules consisting of hundreds to millions of atoms linked together by chemical bonds (typically covalent bonds) which are called primary bonds. The special properties of polymers result particularly from secondary bonds acting between the macromolecules which are known as van der Waals forces. Virtually all types of fibrous material can be used to make nonwoven fabrics depending on:

- the required profile of fabric,
- the cost-effectiveness, and
- the demands for further processing.

The most common fiber types used in nonwoven fabric production are polypropylene, polyester, viscose rayon, polyamide 6 and 6,6, bicomponent fibers, surface-modified fibers, superabsorbent fibers, Novoloid fibers, wool is used in felt production, cotton

in hygienic goods, and another quantitatively important group of fiber raw material for nonwoven is waste fiber materials. The production of nonwovens is carried out after considering the following points:
- Processability in a particular technology
- Impact on the product properties
- Price

Thus, it is very essential to study different fiber properties for the development and production of nonwoven fabrics. According to a study carried out by Tecnon Ltd. the world consumptions of different types of fibers are polypropylene 63%, polyester 23%, viscose rayon 8%, acrylic 2%, polyamide 1.5%, and others 3%. Polypropylene fibers consumption is highest in nonwoven fabric production due to certain properties, e.g., low density, hydrophobicity, low melting and glass transition temperature, biological degradation resistance, chemical stability, and good mechanical strength [2].

5.2 Manufacturing of Nonwoven

The production of nonwoven fabric is comparatively easier as compared to conventional fabric production methods like weaving. First, fiber type is selected from natural or man-made origin, and then selected fibers are converted in the form of a regular sheet or web. In the third step, fibrous sheet or web is bonded together for consolidation and to provide strength of the sheet. Finally, finishes are applied over the consolidated web according to the end use. The production sequence of nonwoven fabric is shown in Fig. 5.2.

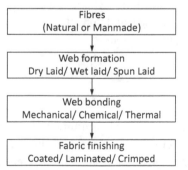

Fig. 5.2: Nonwoven production steps.

Nonwoven fabrics can be classified based on fiber orientation during web formation and bonding of the web. Structure-based classification of nonwoven fabric is shown in Fig. 5.3.

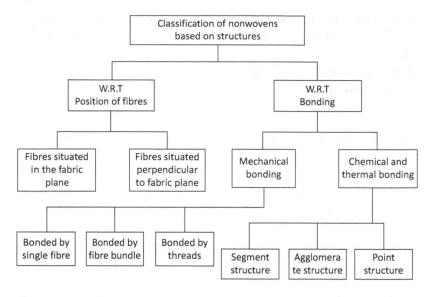

Fig. 5.3: Structure-based classification of nonwoven fabrics.

5.2.1 Web Formation

After the fiber selection, web formation is the major step of nonwoven fabric formation. During web formation fibrous sheet, web, or batt is formed having two-dimensional or three-dimensional assemblies. Orientation, dimension, and structural arrangement of fibers in the web greatly influence the final product properties. Fibrous sheet or web formation is classified into three areas: dry laid, wet laid, and polymer laid (spun-bonded). The dry-laid technique is directly related to the conventional spinning process while wet-laid technique is related to the paper-making industry and polymer-laid technique is directly related to the polymer extrusion through spinneret.

5.2.2 Dry-Laid Web

Dry-laid web formation is concerned with the carding process of spinning. Carding produces one or more webs, in which fibers are preferentially oriented in the machine direction (MD). A multilayer web is produced by using more than one card machine in a single line. The major objective of carding is to disentangle and mix the fibers to form a homogeneous web of uniform mass per unit area. This purpose is achieved by the interaction of fibers with toothed rollers situated throughout the carding machine. The most basic principle of carding is "working" and the second is "stripping." The whole carding process is essentially a succession of "working" and "stripping" actions linked by incidental actions. Every card has a central cylinder or swift that is normally the largest roller and small rollers (called worker and strippers). Generally, small rollers operate in pairs and are located around the cylinder and carry out the basic function of working and stripping. Many cards have more than one cylinder with small rollers. The arrangement of rollers in a basic carding machine is shown in Fig. 5.4.

Worker and stripper pairs around the cylinder perform both opening and mixing functions. Doffer cylinders condense and remove the fibers from cylinder surface in the form of a continuous web. The points of teeth on a worker roller directly oppose the points of cylinder teeth in a point-to-point relationship. The worker revolves in the opposite direction to that of the cylinder. The teeth on worker and cylinder travel in the same lateral direction at their point of interaction, causing a "working" action between the worker and cylinder. Then a stripping action between the stripper and the worker, followed by a further stripping action between the cylinder and the stripper. After removal of the web from the first card machine, it is allowed to pass under the second card machine through a conveyor to achieve the required thickness of the web. Batt drafting is done to increase the fiber orientation in the MD.

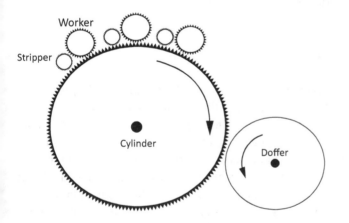

Fig. 5.4: Working and stripping actions.

5.2.3 Wet-Laid Web

The technology of wet-laid nonwovens is closely related to that of paper and paper-making which itself goes back some 2,000 years, developed in China. But wet-laid non-wovens are differentiated from paper manufacturing and regarded as nonwoven: if more than 50% by mass of its fibrous content is made up of fibers (excluding chemically digested vegetable fibers) with a length-to-diameter ratio of greater than 300; or more than 30% by mass of its fibrous content is made up of fibers (excluding chemically digested vegetable fibers) with a length-to-diameter ratio greater than 300 and its density is less than 0.40 g/cm^3.

A dilute slurry of water and fibers is deposited on a moving wire screen and drained to form a web as shown in Fig. 5.5. The web is further dewatered, consolidated, by pressing between rollers, and dried. Impregnation with binders is often included in a later stage of the process. Wet-laid web-forming allows a wide range of fiber orientations ranging from near random to near parallel. The strength of the randomly oriented web is rather similar in all directions in the plane of the fabric. A wide range of natural, mineral, synthetic, and man-made fibers of varying lengths can be used.

Fig. 5.5: Schematic of wet-laid web formation.

5.2.4 Polymer-Laid Web

Polymer-laid, spun-laid, or "spun-melt" nonwoven fabrics are produced by extrusion spinning processes, in which filaments are directly collected to form a web instead of being formed into tows or yarns as in conventional spinning. As these processes eliminate intermediate steps, they provide opportunities for increasing production and cost reductions. Melt-spinning is one of the most cost-efficient methods for producing fabrics. Commercially, the two main polymer-laid processes are spun bonding (spun bond) and melt blowing (melt blown). A primary factor in the production of spun-bonded fabrics is the control of four simultaneous, integrated operations: filament extrusion, drawing, lay down, and bonding. The basic stages of spun-bonded nonwoven fabric include [2]:

Polymer melting → Filtering and extrusion → Drawing → Laydown on forming screen

→ Bonding → Roll up

The first three operations are directly adapted from conventional man-made filament extrusion and constitute the spun or web formation phase of the process, while the last operation is the web consolidation or bond phase of the process. All spun-bond manufacturing processes have two aspects in common:
- They all begin with a polymer resin and end with a finished fabric.
- All spun-bond fabrics are made on an integrated and continuous production line.

Melt-blowing is a process in which, usually, a thermoplastic fiber-forming polymer is extruded through a linear die containing several hundred small orifices. Convergent streams of hot air (exiting from the top and bottom sides of the die nosepiece) rapidly attenuate the extruded polymer streams to form extremely fine diameter fibers (1–5 μm). The attenuated fibers are subsequently blown by high-velocity air onto a collector conveyor, thus forming a fine-fibered self-bonded nonwoven melt-blown web as shown in Fig. 5.6.

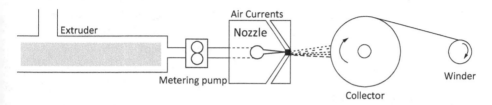

Fig. 5.6: Schematic of melt-blown web formation [2].

In general, high molecular weight and broad molecular weight distribution polymers such as polypropylene, polyester, and polyamide can be processed by spun bonding to produce uniform webs. Medium melt viscosity polymers are commonly used for the production of fibers by melt spinning. In contrast, low molecular weight and relatively narrow molecular weight distribution polymers are preferred for melt blowing. In the past decade, the use of polyolefin, especially polypropylene, has dominated the production of melt-blown and spun-bonded nonwovens. One of the main reasons for the growing use of polyolefin in polymer-laid nonwovens is that the raw materials are relatively inexpensive and available throughout the world.

5.3 Web Bonding

After the web preparation, the next major step in nonwoven production is web bonding. Different web-bonding techniques are available depending upon the end-product properties and cost. Nonwoven web bonding mainly has three categories:
- Thermal bonding
- Chemical bonding
- Mechanical bonding

5.3.1 Thermal Bonding

Thermal bonding requires a thermoplastic component to be present in the form of a fiber, powder, or as a sheath as part of a bicomponent fiber. The heat is applied until the thermoplastic component becomes viscous or melts. The polymer flows due to surface tension and capillary action of fiber-to-fiber crossover points where bonding regions are formed. These bonding regions are fixed by subsequent cooling. No chemical reaction takes place between the binder and the base fiber at the bonding sites. Binder melts and flows into and around fiber crossover points and into the surface crevices of fibers in the vicinity, and adhesive bond is formed by subsequent cooling. Thermal bonded products are relatively soft and bulky depending upon the fiber's composition. Thermal bonding process is economical, environmental friendly, and 100% recycling of fibers components can be achieved. In the thermal bonding technique, generally hot calendar rollers are used to bind the fibrous sheet. Thermal bonding has further subcategories: point, area, infra red, ultrasonic, and through air bonding.

5.3.2 Chemical Bonding

In chemical bonding, different types of chemicals or binders: rubber (latex), synthetic rubber, copolymers, acrylics, vinyl esters, styrene, and different natural resins are sprayed on the nonwoven web for bonding purposes. Latex binder is most commonly used for nonwoven web bonding. During chemical bonding process, either chemicals are sprayed on the nonwoven web or the web is allowed to pass through the chemical box. In chemical bonding, different techniques are used for web bonding. The most frequently used chemical bonding processes are spray adhesives, print bonding, saturation adhesives, discontinuous bonding, and application of powders.

5.3.3 Mechanical Bonding

In the mechanical bonding process, a fibrous sheet or web is bonded together through the application of liquid or air jets, punching needles, and by stitching. Depending upon the selection of any type of mechanical media, nonwovens are classified as hydro entanglement, needle punching, and stitch-bonded fabrics. In the hydro entanglement technique, a fibrous sheet is allowed to pass under the liquid jets provided by multiple nozzles. Through the jet pressure, web is fused, consolidated, and provides strength to the sheet as shown in Fig. 5.7. The major disadvantage of this technique is the drying of the sheet after consolidation. Hydroentangled nonwoven fabrics are used in the wipes and medical nonwoven industry because of their additives and lint-free, soft, strong, and cost-effective characteristics.

While in the needle punching method, a fibrous web is allowed to pass under a bar containing multiple needles. These needles pass through the thickness direction of the web and entangle the fibers to give strength to the fibrous sheet. A schematic of needle punching is shown in Fig. 5.8. Needle-punched nonwovens are used in automotive,

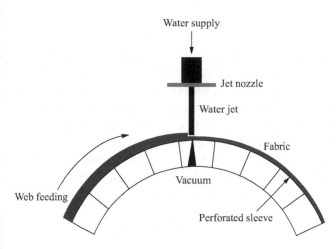

Fig. 5.7: Hydroentanglement bonding assembly [2].

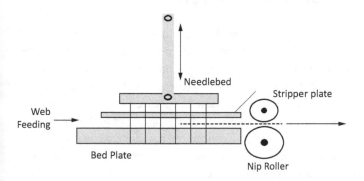

Fig. 5.8: Schematic of needle punching technique [2].

construction, home furnishing industries, geotextiles, shoe felts, blankets, filters, and insulators.

Stitch-bonded nonwoven fabrics are produced by stitching the fibrous web or sheet with other fibers or yarns. The performance properties of stitch-bonded nonwoven fabric depend upon the stitching yarn type, stitch density, stitch length, stitching yarn tension, and machine gauge. Stitch-bonded fabrics may be of one side stitched, two sides stitched, or one side stitched with the projection of pile on the other side of the fabric. To get flexibility in the fabric, Lycra yarn is used and for higher strength fabric, high-performance yarns are used for stitching purposes. Commercially, two stitch bonding systems, Maliwatt and Malivlies, are available. Stitch-bonded fabrics are used to produce vacuum bags, geotextiles, filters, and interlining, and the biggest market is shaped by the home furnishing industry.

5.4 Finishing

Keeping in view the end use of nonwoven fabric, different types of finishes are applied over the fabric. The variety of both chemical and mechanical finishes provides a new horizon for the application of nonwoven fabrics. Different types of wet finishes, dyeing, coating as well as calendaring, embossing, and microcreping were used. These days, many types of chemical finishes like antistatic finish, antimicrobial finish, water-repellent finish, UV absorbers, flame-retardant finish, soil release agent, optical brightener, and super-absorbent finishes are applied on the end product keeping in view the performance application of the product. Plasma treatment, microencapsulation, laser etching, biomimetic, and electrochemical finishes are under developing stages for nonwoven finishing.

5.5 Characterization of Nonwoven

Nonwoven fabric is different from other textile structures because it is produced from fibers or fibrous sheets rather than yarn. In addition to the fiber and binder type, the structural properties of nonwoven fabrics are influenced by the web formation process, bonding, and finishing techniques. The structure and dimensions of nonwoven fabrics are frequently characterized in terms of fabric weight/mass per unit area, thickness, density, fabric uniformity, fabric porosity, pore size and pore size distribution, fiber dimensions, fiber orientation distribution, and bonding segment structure. The majority of nonwoven fabrics have porosities >50% and usually above 80%. The fabric weight uniformity in a nonwoven is normally anisotropic, that is, the uniformity is different in different directions (machine and cross-direction) in the fabric structure. Certain mechanical properties like tensile strength, tear strength, compression recovery, bending and shear rigidity, abrasion, and crease resistance frictional properties are tested according to the end use.

References

[1]   www.INDA.org.
[2]   S. J. Russell, *Handbook of Nonwovens*. Cambridge, England: Woodhead Publishing Limited, 2007.
[3]   www.EDANA.org.
[4]   R. A. Chapman, *Applications of Nonwovens in Technical Textiles*. Cambridge, England: Woodhead Publishing Limited, 2010.
[5]   W. Albrecht, H. Fuchs, and W. Kittelmann, *Nonwoven Fabrics: Raw Materials, Manufacture, Applications, Testing Process*. Weinheim, Germany: WILEY-VCH, 2003.

Waqas Ashraf, Habib Awais, Adeel Abbas

6 Knitted and Braided Fabrics

Abstract: The application areas of fabric are determined by the fabric structure and raw material used. The main categories of fabrics are woven, knitted, braided, and nonwoven fabrics. Knitting is a fabric formation technique in which the yarn is bent into a loop and these loops are interconnected to form the fabric. Knitting can be defined as the interlooping of yarns. The bending of yarn provides better stretch ability, extensibility, comfort, and shape retention properties. However, they tend to be less durable than woven fabrics. Braiding is the fabric formation technique employing the intertwisting or intertwining of yarns in a defined manner. Braiding is also a fabric engineering technique including diagonal interlacement of yarns; however, the angle of interlacement varies from 1° to 89°. Braiding finds wide applications from aesthetics to technical textiles.

Keywords: knitting, warp, weft, circular, braiding

6.1 Knitting

Knitting is the second largest and growing technique of fabric manufacturing in which yarns are interloped to make thick yet flexible and elastic fabric [1]. Knitting is derived from the Dutch word "Knutten" which means to knot. During the Industrial Revolution of the eighteenth century, knitting was primarily performed with machines, and the first knitting machine is thought to be invented by William Lee in 1589. However, this art remained in the hands of underdeveloped and poor sections of society till first half of the twentieth century [2]. Knitting is a fabric formation technique in which the yarn is bent into a loop and the loops are interconnected to form fabric [3]. Knitting can be defined as the interlooping of yarn as shown in Fig. 6.1. The bending of yarn provides better stretchability, extensibility, comfort, and shape retention properties. However, they tend to be less durable than woven.

6.2 Comparison of Woven and Knitted Fabrics

The woven fabrics produced by the interlacement of two sets of yarn and knitted fabrics formed by the interloping of yarn have unique characteristics and have their end-user applications [1]. In most cases, both fabrics can be a substitute for each other, and the selection of the right fabric can meet the requirement of the wearer in a better way. Table 6.1 gives a comparison of woven and knitted fabrics.

https://doi.org/10.1515/9783110799415-006

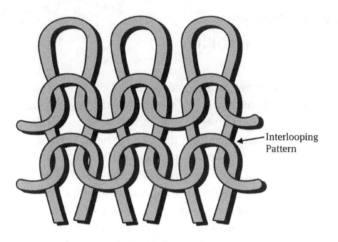

Interlooping
Pattern

Fig. 6.1: Interloping pattern of the knitted fabric.

Tab. 6.1: Comparison of woven and knitted fabric, machine, and process.

S. no.	Parameter	Woven fabric	Knitted fabric
1	Process requirement	The woven fabric requires two sets of yarn for interlacement: one is warp and the other is weft yarn.	Knitting the fabric can be produced from a single end or a cone of yarn in the case of weft knitting.
2	Dimensional stability	More stable	Less stable. Careful handling is required for knitted fabric during wet processing and stitching.
3	Comfort	Less comfort due to tight structure	The more open spaces give better air permeability and moisture management
4	Shape retention properties	Woven garments retain their shape	Knitted garment gets the shape of the wearer's body and is best for undergarments
5	Crease resistant	Poor crease resistance	High crease resistance
6	Development route	Yarn preparation required like warping, sizing drawing, etc.	Fabric can be produced from yarn packages. So process route is very short
7	Conversion cost	The conversion from yarn to fabric involves various processes. The conversion cost is higher as compared to knitted.	The conversion requires no preparation, so the conversion cost is low.

Tab. 6.1 (continued)

S. no.	Parameter	Woven fabric	Knitted fabric
8	Environmental effect	Preparation includes sizing of warp yarn that has to remove before color application, which may cause environmental pollution.	The yarn is just waxed. No need to size the yarn, so development cause fewer environmental hazards.

6.3 Types of Knitting

The knitted fabric can be categorized into two major classes, that is, warp and weft knitting based on yarn feeding and the direction of the fabric formation direction. The weft-knit technique is more common as the fabric can be produced from the single end and there is no need for yarn preparation like warping. In weft knitting, the direction of movement of yarn is in the weft direction of the fabric [1]. The loops are formed horizontally with the same yarn, as shown in Fig. 6.2.

The warp-knitting technique is a more advanced technique and this fabric is much closer to the woven fabric in terms of dimensional stability. The loops that are formed are connected in the warp direction and the movement of yarn is also in the warp direction as shown in Fig. 6.2.

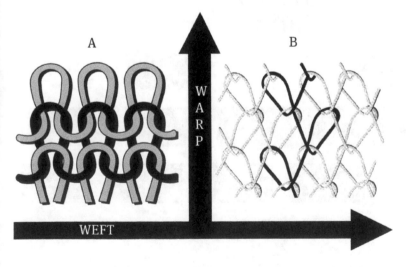

Fig. 6.2: Yarn direction: (A) weft knitting and (B) warp knitting.

6.3.1 Weft Knitting

Weft-knit fabric is familiar for its comfort and shape retention properties. The fabric can be produced from a single end. The movement of yarn in the weft direction provides stretchability in both directions that could be engineered to achieve the required properties. The apparels either inner or outerwear are the most demanding area of weft-knit fabric. This technique is further classified into different machine type and structure that is given in Fig. 6.3. Circular knitting machine is particularly used to produce tubular fabric. Circular machines are classified into three major categories based on cylinder and dial. In the first category in which the machine has only one cylinder, needles are placed inside the cylinder trick that moves up and down for loop formation. Popular structures are single jersey and their derivatives. The second type of machine has both a dial and a cylinder. The needles are placed in both the dial and cylinder. The cylinder needles move up and down while the dial needle moves in a to and fro manner. The major machine types are Rib and Interlock. The difference in their construction is the placement of tricks or grooves. The grooves on the rib machine of both the dial and cylinder are alternative to each other whereas on interlock the grooves are exactly opposite to each other. The third class is purl, in which the machine has two cylinders. These cylinders are superimposed. The purl fabric is also known as link–link fabric. The needle has hocked on both sides of the needle. The same needle is placed in opposite tricks of these cylinders [4]. The flat machines can be produced both single and double knit structures.

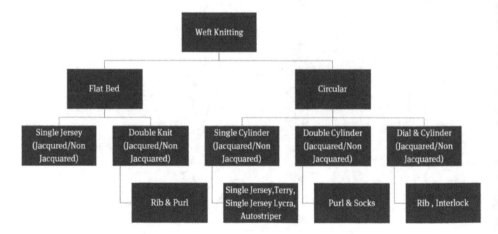

Fig. 6.3: Classification of weft-knitting machines and their structures.

6.3.2 Weft-Knitting Machine Elements

The needle is the most essential part of loop formation. The needles are placed in tricks or cut of bed (flat or circular) at regular intervals so that they can move freely during the loop formation cycle. Most machine manufacturer prefers to use latch needle for their machine. The latch needle is self-actuating, and no auxiliary part is required. Different needle types and their parts are shown in Fig. 6.4.

There are three main types of needles:
1. Latch needle
2. Spring-bearded needle
3. Compound needle

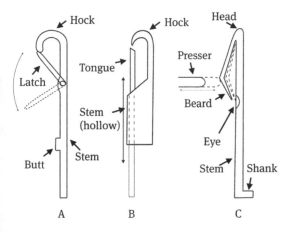

Fig. 6.4: Needle types and their parts: (A) latch needle; (B) compound needle; and (C) bearded needle.

The most important part of the machine after the needle is the sinker. The sinker is a thin metal plate placed on a horizontal surface perpendicular to the needle. They move to and fro in between the needle. The sinkers get their movement from the sinker cams. The purpose of the sinker is to hold the old loop when it is cleared from the needle. The sinker is used both in weft and warp knitting [4]. The different machines which are producing different structures have different types of sinkers used to produce the required results. The machine that has double-bed construction does not need to use a sinker as either bed needle holds the old loop while the other bed needle is in working position. The sinker and its parts are given in Fig. 6.5. The sinker performs the following functions:
– Loop formation
– Holding down
– Knocking over

6.4 Loop Formation Cycle with Latch Needle

The loop formation cycle of the latch needle explains the working of a needle during the loop formation process in weft knitting [5]. The loop formation cycle is given in Fig. 6.6.

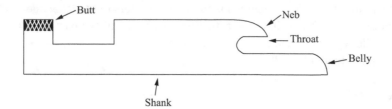

Fig. 6.5: Sinker and its parts.

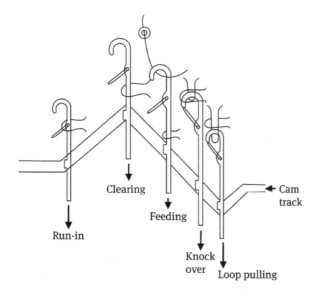

Fig. 6.6: Loop formation cycle of the latch needle.

1. The needle starts moving upward as directed by the cam. The latch opens and the old loop slips on the latch.
2. In this step, the needle clears the old loop and goes to its maximum height.
3. From this step the descending of the needle starts, and the needle engages the new yarn in the hook. This is a feeding of new yarn.
4. The needle moves down. The latch is now closed. This is a knock over position, and the old loop is disengaged from the needle.
5. The last process is the loop-pulling process. The needle goes to the lower-most position and pulls the new loop from the previously formed old loop.

6.5 Principle Stitches in Weft Knitting

There are mainly four basic types of stitches in weft knitting, namely knit, tuck, purl, and miss or float. Mostly weft-knitted fabrics and their derivatives are based on the combination of these stitches.

6.5.1 Knit Stitch

This knit stitch is formed when the needle is raised enough to engage new yarn in the hook by the camming action and the old loop is cleared. The technical back side of the knit stitch is called a purl stitch [6]. The clearing position of the knit stitch is given in Fig. 6.7.

6.5.2 Tuck Stitch

Tuck stitch is formed when the needle is raised to get new yarn but not enough to clear the previously formed old loop. The needle then holds two loops when it descends as shown in Fig. 6.7. The needle can hold up to four loops so it has to clear the previously held old in a wale. The fabric gets thicker with a tuck stitch as compared to knit stitches due to the accumulation of yarn when the needle clears all the held-old loops at the knit stitch. The structure becomes more open and permeable to air than knit stitches. It can also be used to get a different color effect in fabric [4].

6.5.3 Miss or Float Stitch

When the needle does not move upward to clear the old loop and also does not take the new yarn that is presented to it, then a miss or float stitch is formed. The needle is not activated in a miss stitch. Moreover, it holds the old loop as shown in Fig. 6.7. Float stitch on the successive needles produces a longer float of yarn that may cause the problem of snagging. The float is preferably used where we need to hide some color from the technical face of the fabric. The hide yarn floats at the back of the fabric. The yarn gets straighter in float stitch construction, so the extensibility decreases as compared to the tuck and knit stitch [4].

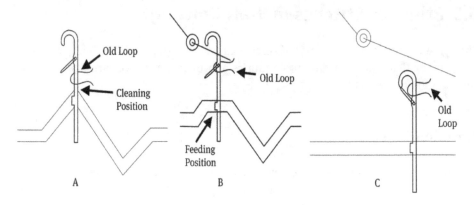

Fig. 6.7: Stitch types: (A) knit; (B) tuck; and (C) float or miss stitch.

6.6 Knitting Terms and Definition

6.6.1 Loop Parts

The needle loop has different parts. The loop parts are important to understand the technical face and back side of the loop. The loop parts are given in Fig. 6.8.

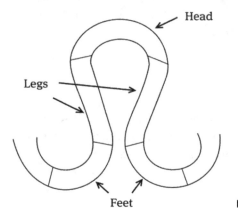

Fig. 6.8: Parts of the knitting loop.

6.6.2 Technical Face and Back

The technical face is the surface of the fabric where the feet of the new loop cross under the legs of the old loop and the legs cross over the head of the old loop. It may also be defined as the side that has all the faces of the knit loop. Figure 6.9 shows the interloping of old and new loops, forming the technical face side. If the feet of the new loop cross over the legs of the old loop and the new loop legs pass under the head of the old loop, then it is said to be a technical back side. The interloping pattern of the technical back is given in Fig. 6.9.

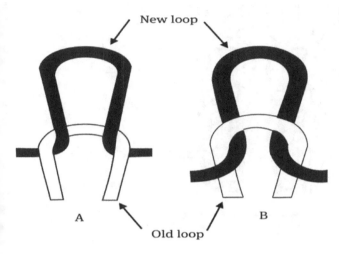

Fig. 6.9: Fabric sides: (A) technical face and (B) technical back.

6.6.3 Wales and Courses

The series of the loop that is meshed vertically are known are wales. The consecutive loops that are connected horizontally are called courses. The rows and columns of loops connected are shown in Fig. 6.10.

6.6.4 Stitch Density

The stitch density of a knitted fabric is expressed in terms of wale density and course density. The number of wales per unit length is called the wales density, while the number of courses per unit length is called the course density. The unit length is normally 1 in. or 1 cm. The knitted fabric structure given in Fig. 6.10 shows three wales in 1 in. and two courses in 1 in. of fabric. The total number of loops or stitches per unit area is called stitch density. It is the product of course and wale density. The stitch density of knitted fabric shown in Fig. 6.10 is 6.

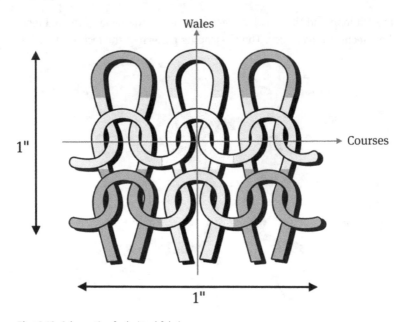

Fig. 6.10: Schematic of a knitted fabric.

6.6.5 Stitch Length

The stitch length is the most important parameter of a knitted fabric. It is the length of yarn consumed to make one complete loop [2]. The knitted fabric's dimensional, physical, and mechanical properties are determined by the stitch length, which can be engineered to meet the requirement of the fabric.

6.7 Warp Knitting

Warp knitting may be defined as the loop formation process along the warp direction of the fabric [6]. The simultaneous sheet of yarn is provided to the machine along the warp direction for the loop formation process. The sheets of yarn are supplied from warp beam like in weaving. Each warp end is provided to an individual needle. The same yarn runs along the warp direction and the needle draws the new loop yarn through the old loop that was formed by another yarn in the previous knitting cycle. Each yarn also passes through the guide mounted on the guide bar which provides the movement of the same yarn between needles.

The warp-knitting machines are flat-bed and the fabric formation technique is more complex as compared with weft knitting. The flow process of warp knitting is given in Fig. 6.11, while a comparison of warp and weft knitting is given in Tab. 6.2.

Yarn Cone/Package 〉 Warping 〉 Warp Knitting Machine

Fig. 6.11: Flow process of warp knitting.

Tab. 6.2: Comparison of warp and weft knitting [4].

S. no.	Weft Knitting	Warp Knitting
1	Individual yarn is provided to the feeder. If a machine has 90 feeders and all are active, then 90 courses are inserted in a complete revolution.	A simultaneous sheet of yarns is fed to the machine. No. of ends required will be equal to the no. of needles on the machines.
2	Loop formation along the weft or course direction of the fabric.	Loop formation along the wale or warp direction of the fabric.
3	The yarn is supplied in the form of cones or cheese. The number of cones required will be equal to the number of feeders available on the machine.	Yarn is supplied to the machine from the warp beam, so an additional warping process is required to prepare the warp beam.
4	The spun yarn is mostly the raw materials for weft knitting so only waxing may require to avoid abrasion between yarn and machine parts.	The filament (synthetic) yarn is used in warp-knitted fabric according to the end application. Antistatic oiling is required to avoid static charge.
5	The weft-knitted fabric has less dimensional stability so careful handling is required.	The warp-knit fabric is dimensionally very stable due to the overlapping and underlapping of yarn.
6	The weft-knit fabric is more stretchable in both directions (warp and weft).	Warp-knit fabrics have lower stretchability in both directions. However, it is comparatively higher in width direction.
7	A latch needle is preferably used in the weft-knitting machine.	All three types of needles (latch, bearded, and compound) are used in a warp-knitting machines. Raschel machines are designed to run latch needles, whereas bearded and compound needles are utilized in Tricot machines.
8	The application of weft-knitted fabric is mostly in apparel including both outer and inner garments.	The application area of warp-knit fabric is not only apparel but also has a huge demand for domestic, industrial, and technical application.

6.7.1 Classification of Warp-Knitting Machine

Warp-knitting machine is categorized based on the construction of different machine parts and their operation. Tricot and Raschel are two main categories of the machine. The further classification of warp knitting is given in Fig. 6.12, while a comparison of Tricot and Raschel warp knitting is given in Tab. 6.3.

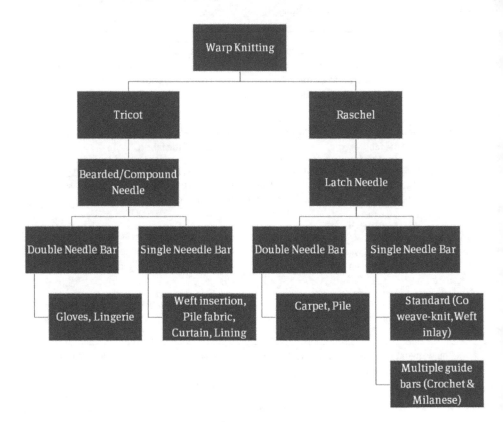

Fig. 6.12: Classification of warp-knitting machines and their structures.

6.8 Applications of Knitted Fabrics

The application area of knitted fabric is mainly classified into three major categories such as clothing which includes weft-knitted vests, sweaters, pullovers, stockings, sportswear, and underwear. The home and furnishing textile is the second major class comprised warp-knitted curtains, terry towels, and weft-knitted blankets, upholstery, etc. Knitted fabrics also have a huge application range in technical textiles. Both warp and weft-knitted fabric are used in medical textiles such as compression bandages. The

Tab. 6.3: Comparison of Tricot and Raschel machine [6].

S. no.	Machine parts and their operations	Tricot	Raschel
1	Needle type	The bearded and compound needle is mostly used.	A latch needle is preferably used in the Raschel machine.
2	Sinkers	Sinkers works throughout the knitting cycle.	Sinkers are active only when the needle rises.
3	Machine speed	The machine can run at a speed of 3,500 courses/min.	Raschel machine works at a bit lower speed of up to 2,000 courses/min.
4	Gauge	Tricot machine gauge is expressed by the number of needles per unit inch.	Gauge defined by the no of needles per 2 in.
5	Warp beams/guide bars	Option to use less no of guide bars that can be up to 8.	A minimum of 2 and a maximum of 78 guide bars can be used.
6	Structure type	The comparatively less complex or simple structure can be developed.	The advanced and complex structure can be produced on Raschel.
7	Take-down angle	The take-down angle is 90°.	The angle is 160°.
8	Fabric width	Wider fabric can be achieved up to 170 in.	The width of the fabric is limited to 100 in.

automobile industry also has the consumption of warp-knitted fabric in the form of seat covers, roofing, and filtration. Packaging materials and mosquito nets are also made with knitted fabric [7].

6.9 Braided Fabrics

Braiding is a simple and more interesting fabric engineering technique used in a broad spectrum of products. It is a fabric formation technique through the intertwisting or intertwining of yarns in a defined manner. Yarn being an intermediate textile product has the capability of being converted into the fabric from several techniques including weaving, knitting, and braiding. Previously discussed woven fabrics follow interlacement patterns of yarns. Braiding is also a fabric engineering technique including diagonal interlacement of yarns; however, the angle of interlacement is not 90°. The angle of interlacement could be from 1° to 89° but usually ranges from 30° to 80° depending on the number of yarns being utilized for braiding [8]. Females' hair braids are the simplest daily life example of braiding and are known as simple braid patterns as braiding is possible using a minimum of three yarns. Braiding utilizes the set of yarns in the longest dimension of fabrics and finds wide applications from aesthetics to technical textiles. Viewing the machine architectures and braiding structure, braids have two major classifications.

6.9.1 Flat-Braided Fabrics

Three-yarn braiding can be further extended to an increased number of yarns for flat braids formation. Mostly five to seven braiding yarns are utilized. Edge yarns (first and last yarn) are not bonded to each other in flat braids creating a tape-like appearance [9]. Flat braiding can also be performed by hand using more than three yarns. The yarns are divided into left and right pairs. A two-step braiding technique is implemented, that is, in the first step left yarns overlap with right yarns and right yarns overlap on left in the second step as shown in Fig. 6.13. Similar principle has been implemented in flat-braiding machines to engineer braided structures having superior quality than hand braids. Flat braiding is possible with both even and the odd number of yarns. A flat braid with an even number of yarns has the exact right and left yarn pairs; however, in case of an odd number of yarns, one yarn remains free creating a float in structure.

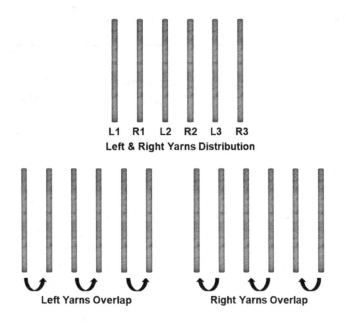

Fig. 6.13: Diagrammatic representation of flat braid.

6.9.2 Tubular-Braided Fabrics

The tubular-braiding technique architects the structures having edge yarns (first and last yarn) bonded to each other. Such bonding creates a hollow tube-like appearance; hence the braids are called tubular braids. Tubular braiding is possible using an even number of yarns because of spool pairing on the braiding machine [10]. Like flat braiding, the yarns overlapping phenomena are followed for tubular braids formation, but the yarns are placed around a circular platform. A simple example of tubular braids is the eight yarns plain braid.

For understanding let us recall the left and right pairs of flat braiding; here instead of left and right the yarns are numbered from one to eight. In flat braiding left and right pairs remain the same in a plain braided structure. However, tubular braiding pairs become an alternative in a two-step process of tubular braiding [4]. Initially, the pairing sequence is (1–2, 3–4, 5–6, and 7–8) as shown in Fig. 6.14(a), and the first overlap of yarns occurs in the pairs. After the first overlap, the pairs are changed with (1–8, 6–7, 5–5, and 2–3) sequence and the second overlap of yarns occurs (Fig. 6.14(b)).

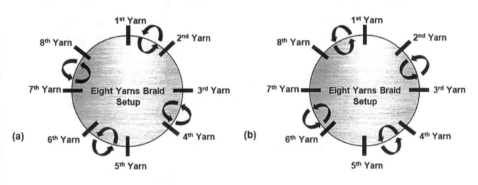

Fig. 6.14: Eight yarns tubular braid.

6.10 Maypole Dance

Maypole is a European traditional dance in which the dancers revolve around a pole in a defined sequence while holding a rope or fabric in each dancer's hand. Half of the dancers walk in the clockwise direction and half in the anticlockwise direction [11]. The top ends of the yarns are fixed on a pole and the pole is slowly covered with a braid from the top toward the bottom as the dancers proceed. The central pole is also called a core and is also termed a mandrel if is of a complex shape. Inlaid tubular braids (hollow tubular braids consisting of a straight yarn passing through the core) have originated from the maypole dance.

6.11 Braiding Machine Setups

Different braiding machine setups are available in industries depending on the end requirements of users. Usually, major structural design variations of braids are a precursor for changing machine setups. Significantly used braiding machines and their basic working principles have been explained in this section.

6.11.1 Maypole Braiding Machine

Maypole dance principle-oriented braiding machines utilize a circular platform on which spools are placed in the form of pairs. Spools are the yarn storage devices onto which yarn is wound from respective bobbins for being utilized on braiding machines [12]. Spools are smaller in size than conventionally used cones and have flanged edges. A special spool-winding machine is utilized for yarn winding onto spools. The spool winding is considered and only preparatory process in braided fabrics engineering. Spools are placed onto spool holders mounted on the circular platform of the maypole braiding machine. The bottom of the spool holders is pinned inside the circular tracks which provide the path for spools movement. Likewise, in maypole dancers, half of the spools move in the clockwise direction and half anticlockwise (Fig. 6.15) [13].

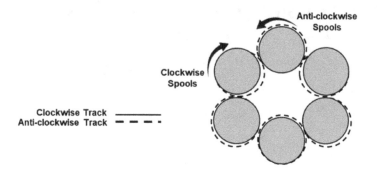

Fig. 6.15: Clockwise and anticlockwise braiding tracks.

Auxiliary machine elements are necessary for smooth braiding operations. Horn gears are mounted below the round platforms comprising horn gear slots in which spool holders are present. Spool holders provide safe spool handling and yarn unraveling. The holders are also termed spool carriers. A spool carrier is a complex assembly of mechanical parts. It comprises a stick to mount the spool, and the spool is free to rotate over there. Negative yarn feeding toward the braiding zone automatically rotates the spool during unwinding. Yarn from the spool is passed through a close guide tensioner mounted above as shown in Fig. 6.16. Mechanical stop motion guide remains in the air due to passing yarn tension; however when the yarn breaks it falls touching the surface of the horn gear, electrical circuit completes, and the machine stops. Another top close guide is present on the carrier to project the yarn toward the braiding zone. Guides and tensioners integrated into the carrier not only provide yarn with a specified passage of movement but also prevent yarn entanglements with other working parts of the machine. Flat braids are also engineered on the maypole principle machines where the spools do not complete their rotation. After 180° rotation both spools reverse their motion and move again toward a zero-degree position. Such half revolutions create a flat structure instead of the tube.

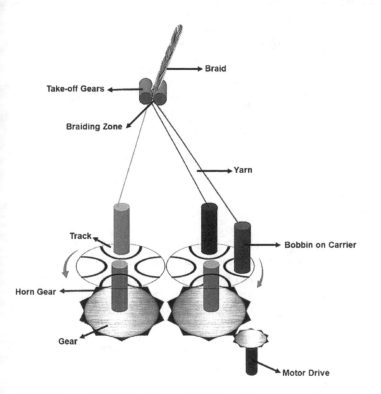

Fig. 6.16: Maypole tubular braiding machine elements.

6.11.2 Three-Dimensional Braiding/Form Braiding

Specialty braided products require three-dimensional (3D) shapes instead of flat and tubular architectures. Hence a need arises for 3D-braided geometries with rectangular, hexagonal, or any other required shapes. 3D braids are thicker as compared to normal tubular braids and can be easily segregated [14]. 3D braids have a greater number of spool tracks compared to conventional braiding machines as shown in Fig. 6.17. There is another terminology called diagonals. Braiding machines with three diagonal tracks are named 3D and 4D for four diagonal tracks [15]. A square-shaped 3D structure can be engineered using four diagonal tracks. In terms of dimensions, the structure will be 3D but for diagonals, it will be a 4D-braided structure [16].

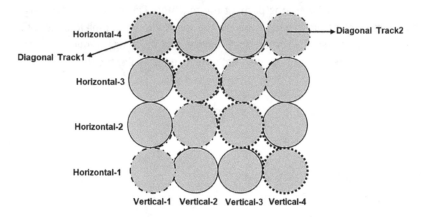

Fig. 6.17: Four-track 3D-braiding platform.

6.11.3 Lace Braiding

Lace-braiding machines comprise a flat platform with round spool track plates. Horn gears are mounted below the round track plates. Spools are arranged according to column numbers. Spools shift their column numbers with respect to required lace patterns. The simplest lace braid can be engineered using two round-track plates with two spools placed on each plate. The spools shift their column numbers and a lace pattern is formed.

6.12 Application Areas

Braided fabrics find wide applications in both conventional and technical textiles. Shoe-laces are the significantly seen tubular and flat-braided products. Braided laces are utilized for aesthetic purposes in women's clothing. Although warp-knitted laces are more common, braided laces still hold their place. Home furnishings require braids as trimming materials for raw edges of fabrics. Surgical sutures in medical textiles are engineered through the braiding technique. Electrical wires cover, ropes, and industrial belts are also braided. The 3D-braided structures provide a low-cost alternative for complex shapes and geometrical architecting. Apart from the discussed areas braided fabrics have been chiefly used in several areas, which will be hard to summarize in a nutshell.

References

[1]   D. J. Spencer, *Knitting Technology A Comprehensive Handbook and Practical Guide*. 2001.

[2]   T. Gries, D. Veit, and B. Wulfhorst, *Textile Technology*. Henser Publications, 2006.

[3]   T. Hussain, State of textile and clothing exports from Pakistan, TEXtalks, vol. July/ August, pp. 62–65, 2013.

[4]   D. Anbunami, *Knitting Fundamentals, Machine Structure and Development*. New Delhi: New Age International (P) Limited, 2005.

[5]   Y. Nawab, T. Hamdani, and K. Shaker, Eds., *Structural Textile Design: Interlacing and Interlooping*. Boca Raton: CRC Press, 2017.

[6]   S. C. Ray, *Fundamentals and Advances in Knitting Technology*. 2011. doi: 10.1533/9780857095558.

[7]   K. F. Au, Eds., *Advances in Knitting Technology*. Woodhead Publishing, 2011. doi: 10.1533/9780857090621.3.287.

[8]   Y. Kyosev, *Braiding Technology for Textiles*. 2015. doi: 10.1016/c2013-0-16172-7.

[9]   Y. Kyosev, "Topology based models of tubular and flat braided structures," Topology-Based Modeling of Textile Structures and Their Joint Assemblies, pp. 13–35, 2019. doi: 10.1007/978-3-030-02541-0_2.

[10]   Y. Kyosev, Generalized geometric modeling of tubular and flat braided structures with arbitrary floating length and multiple filaments, Textile Research Journal, vol. 86, no. 12, pp. 1270–1279, Jul. 2016. doi: 10.1177/0040517515609261.

[11]   S. Xiao, P. Wang, D. Soulat, and X. Legrand, "Structure and mechanics of braided fabrics," Structure and Mechanics of Textile Fibre Assemblies, pp. 217–263, Jan. 2019. doi: 10.1016/B978-0-08-102619-9.00007-9.

[12]   K. Bilisik, "Braiding and recent developments," Fibres to Smart Textiles, pp. 131–152, Aug. 2019. doi: 10.1201/9780429446511-7.

[13]   D. Branscomb, D. Beale, and R. Broughton, New directions in braiding, Journal of Engineered Fibers and Fabrics, vol. 8, no. 2, 2013.

[14]   X. Li, et al., Research status of 3D braiding technology, Applied Composite Materials, vol. 29, no. 1, pp. 147–157, Feb. 2022. doi: 10.1007/S10443-021-09963-2/FIGURES/7.

[15]   Y. Kyosev, *Advances in Braiding Technology*. 2016. doi: 10.1016/c2014-0-03190-5.

[16]   C. Emonts, et al., Innovation in 3D braiding technology and its applications, Textiles, vol. 1, no. 2, pp. 185–205, Jul. 2021. doi: 10.3390/TEXTILES1020009.

Tanveer Hussain

7 Textile Processing

Abstract: Textile processing involves a group of processes including preparation, coloration (i.e., dyeing or printing), and finishing. Depending upon the form of textile material, the fiber, or blend in fabric, or the end use, the process flow chart may have some variations. A fabric may be dyed or printed either after bleaching or mercerizing, depending upon the requirements.

Keywords: pretreatments, dyeing, printing, finishing

7.1 Introduction

Textile materials in different forms, such as fiber, yarn, woven fabric, knitted fabric, or garment may be subjected to different textile processing operations. A general textile processing flowchart is given in Fig. 7.1. An additional heat-setting process may be required for fabrics containing synthetic fibers. Some processes may be combined, for example, scouring or bleaching may be combined in one operation; similarly, dyeing may be combined with some chemical finishes or finishing may be done directly after bleaching. These processes are briefly described in the following sections, taking the example of woven fabrics.

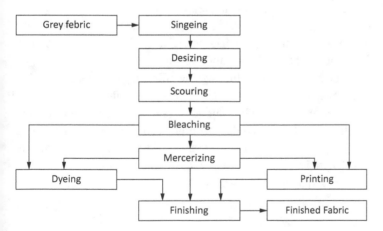

Fig. 7.1: General steps in textile processing.

https://doi.org/10.1515/9783110799415-007

7.2 Textile Pretreatments

The main objective of fabric preparation or pretreatments is to remove any impurities or contaminants from the fabric and make it ready for subsequent operations such as optical brightening, dyeing, printing, or finishing. The overall objectives of fabric preparation include the removal of fabric impurities such as protruding fibers, sizing agents, cotton seed husks, fats, oils, waxes, dirt, dust, and lubricants.; improved absorbency and/or whiteness of the fabric; minimum fiber damage; uniform residual size, pH, alkalinity, whiteness, absorbency, and moisture content. The commonly used fabric pretreatment processes are briefly described in the following sections.

7.3 Singeing

Singeing is a process of passing an open-width fabric over a gas flame at such a distance and speed that it burns only the protruding fibers but does not damage the main fabric. The main objective of the singeing process is to produce a clean fabric surface and reduce fabric pilling tendency by removing the protruding fibers from the fabric surface.

There are three different methods of singeing such as hot plate singeing, hot rotary cylinder singeing, and gas flame singeing. However, gas flame singeing is the most commonly used singeing method because of better-singeing efficiency. Different singing positions may be used for different fabrics to attain the different intensities of singeing. The main singeing positions are: singeing onto free-guided rollers; singeing onto water-cooled rollers, and tangential singeing [2].

The main parameters of gas flame singeing include gas flame intensity, fabric speed over the flame, singeing position (or angle between the fabric and the flame), the distance between the fabric and the flame, and the width of the flame.

The commonly used methods to assess the singeing efficiency include observation of fabric surface with a magnifying glass or stereo microscope and fabric pilling test. A harsher fabric hand-feel may also be an indication of oversingeing. Commonly encountered problems in singeing operations include: incomplete singeing, uneven singeing along the fabric length, uneven singeing along the fabric width, horizontal/width-way singeing stripes, vertical/length-way singeing stripes, over-singeing or thermal damage to the fabric, and formation of molten fiber beads in synthetic or blended fabrics.

7.4 Desizing

Desizing is a process of removing sizing agents from the fabrics, which are usually applied to the warp yarns before weaving. Sizing agents mostly comprise macromolecular film-forming and fiber-bonding substances such as starch, PVA, and polyacrylates, which are applied on warp yarns to increase their strength and reduce yarn breakages during weaving. Some auxiliaries such as wetting agents, softening agents, lubricating agents, or hygroscopic agents may also be included in the sizing recipe.

Removal of sizing agents after weaving is necessary in order to make the fabric more absorbent for dyes and other chemical processing agents. The choice of method for desizing depends on the type of sizing agents used during sizing. Different desizing methods include enzymatic desizing, oxidative desizing, acid desizing, desizing with hot alkali treatment, and desizing with hot detergent solutions. Starch-based sizing materials are usually not water-soluble and require the use of amylases enzymes for their removal. The enzymatic method is most commonly used for desizing fabrics that contain starch-based sizing materials. Advantages of enzymatic desizing in comparison with oxidative and acid desizing include no damage to the fiber, no use of aggressive chemicals, high biodegradability, and a wide variety of application methods.

Water-soluble sizing materials such as polyvinyl alcohol (PVA) or carboxymethyl cellulose (CMC) may be removed by simply washing them with hot water without the use of any enzymes. Important parameters for the enzymatic desizing method include liquor pH, temperature, time, type, and amount of any wetting agent or detergent. The effectiveness of the desizing process can be checked by putting a drop of iodine solution on the desized fabric, which turns blue color if there is still some unremoved starch present in the fabric [3].

7.5 Scouring

Scouring is a process for removing natural and acquired impurities from fabrics to make them more absorbent and suitable for subsequent processes such as bleaching, dyeing, printing, or finishing. Natural cotton contains a very small amount of oils, fats, and waxes, in addition to acquiring dirt or dust, during transportation and storage, which make it dirty and less absorbent. Cotton fabrics are usually scoured by using liquors containing strong alkalis such as caustic soda and detergents at boiling temperature. Impurities such as calcium or magnesium may also be present not only in the cotton fiber but also in the processed water. Sequestering agents may also be used during scouring to counter the negative effect of calcium and magnesium on scouring. Bioscouring is an alternative method to conventional scouring with caustic soda and detergents, in which enzymes such as pectinases are used for cotton scouring.

Fabrics made from wool are also subjected to scouring. However, since wool is sensitive to alkalis, milder scouring conditions are used for wool. Since fabrics made from regenerated and synthetic fibers do not only contain any natural impurities but only small amounts of processing lubricants and dirt, their souring also requires milder conditions as compared to those required for cotton fabrics. Since scouring is a process for removing natural and acquired impurities from fabrics, the severity of the scouring treatment depends upon the nature and amount of the impurities present as well as the sensitivity of the fiber to different scouring agents. The aim is to remove the impurities without damaging the fiber.

The main parameters of the scouring process include type and amount of alkali, type and amount of detergent, type and amount of sequestering agent, temperature, and time of the treatment. In the case of enzymatic scouring, the main parameters include the type and amount of enzymes, pH, temperature, and time of the treatment. The efficiency of scouring treatment can be evaluated by testing fabric absorbency and determining fabric weight loss after the treatment. A cotton fabric that can absorb a drop of water within a few seconds is considered to be well-scoured [1, 3].

7.6 Bleaching

The purpose of bleaching is to remove any coloring matter from the fabric and confer it a whiter appearance. In addition, to increase fabric whiteness, the bleaching process may also result in improved fabric absorbency and removal of cotton seed husks and trash from the fabric. Although there are different bleaching agents which may be used for bleaching textile fabrics, such as sodium hypochlorite or calcium hypochlorite, but hydrogen peroxide is the most commonly used bleaching agent for cotton and its blended fabrics because of its advantages in comparison with other bleaching agents.

The main parameters for bleaching with hydrogen peroxide include the concentration of hydrogen peroxide, the concentration of alkali (e.g., NaOH), type and concentration of bleaching stabilizer (e.g., sodium silicate), type and concentration of the sequestering agent, pH, temperature, and treatment time. A pH in the range of 10.2–10.7 is considered optimum for bleaching cotton fabrics with hydrogen peroxide. The quality of the bleached fabric may be evaluated by testing its whiteness with the help of a spectrophotometer, fabric absorbency, and checking for any fabric damage [1, 3].

7.7 Mercerizing

Cotton and its blended fabrics are sometimes subjected to a mercerization process to enhance various properties such as an increase in dye affinity, chemical reactivity, dimensional stability, tensile strength, luster, and fabric smoothness. The mercerization process is performed using caustic soda. Although caustic soda is also used in scouring cotton fabrics the concentration of caustic soda is very low in scouring (e.g., 5–10%) while the concentration in mercerization may be up to 22–25%. The main mercerization parameters include the concentration of caustic soda, type and concentration of the wetting agent, temperature, fabric liquor pick-up, time duration, and tension on fabric during the process. Fabric pick-up in mercerization is usually kept as near to 100% as possible and the dwell/contact time with caustic liquor is 40–60 s. After mercerization, the fabric is rinsed for the removal of alkali and its pH is neutralized to make it suitable for subsequent processing. The mercerization process is only done for those fabrics which are made from 100% cotton or contain a substantial amount of cotton in case of a blend. Mercerization is not done for purely synthetic fabrics such as those made from polyester or nylon [1, 3].

7.8 Heat Setting

The heat-setting process is only used for synthetic fabrics such as those made from polyester or their blends to make them dimensionally stable against subsequent hot processes. Other benefits of heat setting include less fabric wrinkling, low fabric shrinkage, and reduced pilling tendency. The heat-setting process involves subjecting the fabric to dry hot air or steam heating for a few minutes followed by cooling. The temperature of heat-setting is usually set above the glass transition temperature and below the melting temperature of the material comprising the fabric [4].

7.9 Textile Dyeing

Two major processes used for the coloration of textiles are dyeing and printing. Dyeing can be defined as a process during which a textile substrate is brought in contact with the solution or dispersion of a colorant, and the substrate takes up the said colorant with reasonable resistance to its release form the substrate. Dyeing comprises the application of colorant to the entire body of a textile substrate with a reasonable degree of fastness. Textile materials can be dyed in fiber, yarn, fabric, or garment form. The dyeing of fibers is known as "stock dyeing." The addition of colorant to the polymer melt o solution before their extrusion is called "dope dyeing" or "solution dyeing." Dyeing of yarns in the form of wound packages, skeins, or beams is known as "package dyeing," "skein dyeing," or "beam dyeing," respectively. Fabric dyeing is also known as "piece dyeing" [5].

The different requirements that a dyer has to meet include: matching the required shade on the dyed material; achieving level/uniform dyeing; obtaining the required degree of color fastness (i.e., resistance to washing, rubbing, light, perspiration, etc.); avoiding any deterioration of textile properties during dyeing (e.g., loss in strength and softness); keeping the dyeing cost as low as possible; and minimizing the harmful impact on the environment.

7.10 Dyes and Pigments

There are two main types of colorants: pigments and dyes. Dyes are either soluble in the dyeing medium (e.g., water) or can dissolve into the textile substrate. Pigments are neither soluble in the dyeing medium nor can dissolve into the substrate. Both pigments and dyes can be natural or synthetic. Colorants from natural sources such as plants have been obtained since prehistoric times. The first synthetic dye was accidentally discovered by an English chemist named William Henry Perkin in 1856.

From the application point of view, dyes have been classified into different groups; each group is suitable for certain types of textile substrates. Some commonly used dyes and their suitability for different fibers are given in Tab. 7.1. For example, the most commonly used type of dye for cotton, polyester, and acrylic are reactive dyes, disperse dyes, and basic dyes, respectively.

Tab. 7.1: Main dye classes and their suitability for different fibers.

Fibers	Class of dyes							
	Acid	Basic	Direct	Disperse	Azoic	Reactive	Sulfur	Vat
Cotton			☑		☑	☑	☑	☑
Viscose			☑		☑	☑	☑	☑
Tencel®			☑		☑	☑	☑	☑
Modal®			☑		☑	☑	☑	☑
Bamboo			☑		☑	☑	☑	☑
Flax			☑		☑	☑	☑	☑
Ramie			☑		☑	☑	☑	☑
Wool	☑					☑		
Silk	☑					☑		
Acrylic		☑		☑				
Nylon	☑			☑				
Polyester				☑				
Triacetate				☑				

7.10.1 Acid Dyes

The acid dyes are usually applied under acidic conditions. They are commonly used for dyeing protein fibers (e.g., wool and silk) and nylon fibers. Acid dyes are anionic, and their negatively charged anions are attracted by positively charged amino groups in wool under acidic conditions. The application of acid dyes on wool or nylon results in ionic bonds or salt links between the anionic dye and the positively charged groups in the fiber under acidic conditions. In addition to ionic bonds, hydrogen bonds as well as Van der Waal's forces may also be formed between the fiber polymer system and the acid dye molecules.

Because of the high dye-fiber affinity due to opposite charges, there is a risk of rushing dye molecules toward the fibers at a high rate with the possibility of unlevel dyeing. To avoid unlevel dyeing, some retardation in the dyeing rate is obtained by making use of sodium sulfate.

The acid dyes are further classified into three main groups:
- Leveling dyes
- Milling dyes
- Super-milling dyes

The main differences in the above three types of dyes include their molecular weight, affinity for fiber, leveling properties, amount of leveling agent required, dyeing pH, and fastness properties. Leveling dyes have the lowest affinity and the best leveling properties but poor wash fastness. Super-milling dyes have the highest affinity, and the worst leveling properties but good wash fastness. The leveling properties of dyes can be improved with careful control of dyeing parameters.

7.10.2 Azoic Dyes

The azoic dyes are named so because of the presence of an azo group in their molecule. They are also known as naphthol dyes. Azoic dyes may be used for dyeing cellulosic materials and some man-made fibers. However, these dyes are not popular these days due to difficulties in their application and shade-matching.

7.10.3 Basic Dyes

The basic dyes are most commonly used for dyeing polyacrylonitrile or acrylic materials. They are also known as cationic dyes because of the presence of a positive charge in the dye molecules under dyeing conditions. During dye application, the negatively charged acrylic fiber attracts the positively charged dye cations for ionic bonding. Due to the high attraction between the oppositely charged fiber and dye molecules, there is a risk of unlevel dyeing because of the high rate of dyeing. This risk may be reduced by careful control of dyeing temperature and the use of suitable retarding agents.

The basic dyes are well known for their intense hues and brilliant shades, unrivaled by any other class of dyes. Basic dyes have excellent light fastness because of their resistance to the destructive effect of ultraviolet radiation in sunlight. Their washing fastness is also quite good, which may be attributed to the hydrophobic nature of the acrylic fiber and the good substantivity of the dye for the fiber.

7.10.4 Direct Dyes

Direct dyes are one of the cheapest groups of dyes used for dyeing cotton and other cellulosic materials. They are water-soluble and can be applied relatively easily using a variety of methods. These dyes are anionic in nature and have a negative charge in an aqueous solution, as do the cellulosic fibers. The addition of common salt (sodium chloride) or Glauber's salt (sodium sulfate) is usually necessary during dyeing to overcome the repulsion between the negatively charged dye and the substrate. After absorption into the fiber, these dyes are held to the fiber by hydrogen bonding and/or van der Waals forces.

Based on their leveling properties, direct dyes are grouped into three main classes:
- Class A: self-leveling dyes
- Class B: dyes with average leveling properties (controlled salt addition improves leveling)
- Class C: dyes with poor leveling properties (controlled salt addition and careful temperature control improves leveling)

Direct dyes usually do not have very good wash fastness properties and tend to fade away from the fabric on repeated washings. However, their fastness properties can be improved by various after-treatments including:
- Treatment with cationic agents
- Treatment with copper sulfate
- Treatment with chrome compounds, such as potassium dichromate
- Treatment with combined potassium dichromate and copper sulfate
- Treatment with formaldehyde

7.10.5 Disperse Dyes

The disperse dyes are mainly used for dyeing polyester. Disperse dyes have extremely low water solubility and are usually used in the form of aqueous dispersions. From an application point of view, disperse dyes can be classified as follows:
- Low-energy disperse dyes have a high rate of diffusion, can be dyed easily with a carrier at atmospheric pressure, and have poor sublimation fastness.
- Medium-energy disperse dyes have moderate diffusion rate, usually require high-temperature exhaust dyeing method, and have moderate sublimation fastness.
- High-energy disperse dyes have a low rate of diffusion, require very high dyeing temperature, and have very good sublimation, wet, and light fastness properties.

The disperse dyes are usually applied in acidic pH in the presence of a dispersing agent. Other dyeing auxiliaries may include a wetting agent, leveling agent, and a dyeing carrier.

7.10.6 Reactive Dyes

The reactive dyes constitute the most commonly used class of dyes for dyeing cellulosic textiles because of their good all-round properties such as water solubility, ease of application, variety of application methods, availability of different shades, brightness of color shades, good to excellent wash and light fastness, and moderate price. Reactive dyes may have poor fastness to chlorine bleach.

The reactive dyes are further classified according to the type of their reactive groups, giving them different degrees of reactivity. For example, dichlorotriazine-based dyes are highly reactive and give good dyeing results at low dyeing temperatures, whereas dyes based on trichloropyrimidine have poor reactivity and give good color yield only at high dyeing temperatures. Vinyl-sulfone-based dyes have moderate reactivity.

The important process variables for dyeing with reactive dyes by exhaust method include dyeing temperature, type, and amount of electrolyte (e.g., common salt or Glauber's salt), dyeing pH (controlled by the type and amount of alkali used), liquor-to-material ratio (L:R), and dyeing time. The fixation of reactive dyes on cellulosic fibers takes place through the formation of covalent bonds under alkaline conditions (pH 9–11).

The typical exhaust dyeing procedure involves the exhaustion of the dye onto the substrate with salt addition and temperature control, followed by the addition of alkali for dye fixation through covalent bonding. After the dyeing process, any unfixed dye or hydrolyzed dye (i.e., the dye which has reacted with water in the dyebath instead of the cellulose) is removed by washing it off using a suitable detergent.

7.10.7 Sulfur Dyes

The sulfur dyes are named so because of the presence of sulfur atoms in their molecules. Like direct dyes, sulfur dyes are also quite cheap for dyeing cellulosic textiles with limited color fastness properties. Different types of sulfur dyes include:
- CI sulfur dyes
- CI leuco sulfur dyes
- CI solubilized sulfur dyes
- CI condensed sulfur dyes

The commonly used sulfur dyes are not soluble in water and need to be converted into a soluble form by reduction with the help of a reducing agent and an alkali. Sulfur dyes are usually easier to reduce and more difficult to re-oxidize as compared to vat dyes. General phases in the dyeing of cellulosic materials with sulfur dyes include:
- Reduction: conversion of the water-insoluble dye into a soluble form in the presence of a reducing agent and alkali
- Application: absorption of solubilized sulfur dye by the textile substrate

- Rinsing: removal of any loose color from the substrate before oxidation
- Oxidation: conversion of the dye absorbed by the textile substrate back into the insoluble form
- Soaping: for increased color brightness and fastness of the final shade

Sulfur dyes have a fair degree of light fastness due to poor stability of the dye molecule to ultraviolet radiations present in sunlight which degrade the dye chromophore. The washing fastness of sulfur dyes is poor because of the inherent limitations of the sulfur dye chromophores. However, like direct dyes, the fastness properties of sulfur dyes can be improved with suitable after-treatment of the dyed textile materials. Unlike direct dyes, the color range of sulfur dyes is quite limited to black, brown, olive, and blue shades. Moreover, sulfur dyes are not available in bright colors as they are available in other classes of dyes. However, sulfur dyes are quite cheap and may be economical in dyeing deep black and navy shades.

7.10.8 Vat Dyes

The name "vat" comes from the wooden vessel which was first used for the reduction and application of vat dyes. Vat dyes are among the most expensive dyes used for dyeing cellulosic materials with the best overall fastness properties, including washing fastness, light fastness, and chlorine fastness. They are preferred for dyeing workwear or uniforms or where the textiles and apparel are expected to undergo repeated industrial laundering.

The vat dyes are generally not soluble in water. However, solubilized vat dyes are also available but are usually more expensive as compared to generally available insoluble vat dyes. Based on their chemistry, vat dyes can be classified into two main groups: indigo derivatives and anthraquinone derivatives. In general, the fastness properties of anthraquinone-based vat dyes are usually better as compared to those of indigo-based dyes. Indigo blue vat dyes are commonly used for producing indigo denim, with different wash-down and worn-out looks. Based on application properties, vat dyes are classified into four main types: IN vat dyes, IW vat dyes, IK vat dyes, and IN special dyes. Major differences in the above four groups of vat dyes include their leveling properties, dyeing temperature, and amount of alkali, salt, and leveling agent required during dyeing. The general phases in dyeing with vat dyes are as follows:

- Reduction: conversion of insoluble vat dye into soluble sodium leuco-vat anions, with the help of a reducing agent (sodium dithionite) and alkali (sodium hydroxide).
- Diffusion: penetration of the reduced/solubilized sodium leuco-vat anions into fibers.
- Rinsing: removal of excess alkali and the reducing agent from the dyed material.
- Oxidation: conversion of vat dye absorbed in the fibers back into an insoluble form.

– Soaping: during which the vat dye molecules absorbed by the textile material are reorientated and associate into a more crystalline form.

The vat dyes have excellent light fastness due to stable electron arrangement in the chromophore (color-bearing group) of the dye molecule and the presence of numerous benzene rings. Vat dyes have excellent wash fastness owing to the aqueous insolubility of the oxidized dye absorbed in the fiber and due to large vat dye molecules trapped within the polymer system. However, vat dyes are usually very expensive and need more expertise for their application because of the greater number of steps involved in dyeing.

7.10.9 Pigments

The pigment colorants usually have no affinity for any type of fiber. They also do not have any ability to form chemical bonds with the fibers. They are commonly applied with the help of chemical binders which keep them adhered or bound to the textile materials. With the help of binders, pigments can be used for dyeing or printing all types of fibers and their blends. Dyeing with pigments usually comprises padding the textile in the pigment and binder dispersion (along with other suitable auxiliaries), followed by drying and curing at a suitable temperature. The pigment-dyed fabrics are usually stiffer (because of the use of binders) and have poor rubbing fastness properties as compared to the fabrics dyed with dyes. However, the pigment dyeing process can be more conveniently combined with the finishing process resulting in more economical and ecological processing.

7.11 Dyeing Methods

Dyeing methods can be classified into two main types: exhaust dyeing and pad dyeing [6].

7.11.1 Exhaust Dyeing

In exhaust dyeing, a finite amount of textile materials (in the form of fibers, yarn, or fabric) is placed in the dye liquor and remains in its contact throughout the dyeing time, during which the dye molecules gradually move (or exhaust) from the liquor toward the fabric for absorption and fixation in the textile material. The rate of dye exhaustion, absorption, and fixation is controlled with the help of dyeing temperature, liquor agitation, pH, or auxiliaries such as electrolytes, alkalis, leveling agents, or retarding agents. The L:R is also an important factor in exhaust dyeing, that is, the

ratio between the amount of liquor and the weight of textile material dyed in that liquor in a batch. The total dyeing time required in exhaust dyeing depends on several factors including depth of shade, type of dyestuff, nature of textile material, and type of dyeing machine. The general phases in exhaust dyeing include the following:
- Disaggregation of dye particles in an aqueous solution or dispersion
- Exhaustion or movement of the dye molecules from the solution/dispersion toward the textile substrate
- Adsorption of the dye molecules on the surface of the textile substrate
- Absorption, penetration, or diffusion of the dye molecules into the fibers of the textile substrate
- Fixation of the diffused dye in the fibers through chemical bonding or by some other mechanism

7.11.2 Pad Dyeing

In the pad-dyeing method, a continuous batch of fabric in open width passes through an impregnator (or padding trough) containing dye liquor, followed by a passage between a pair of squeeze rollers. The pressure of the squeeze rollers can be adjusted to obtain a desired wet pick-up. For example, a wet pick-up of 100% would result in fabric twice its original dry weight after the impregnation and squeezing. The concentration of the dye in the padding tough and the wet pick influences the final depth of color obtained on the fabric. After passing through the squeeze rollers, the fixation of the dye on the fabric may be accomplished by a variety of means including making a batch of fabric and keep rolling the batch for a specific period (pad-batch dyeing method); passing the fabric through a drying and fixation unit (pad-dry-fix dyeing method); passing the fabric through a drying and steaming unit (pad-dry-steam dyeing method); passing the fabric through a steaming unit (pad-steam dyeing method). After both the exhaust and pad dyeing methods, the dyed fabric is usually subjected to a washing/rinsing step to remove any unfixed dye from the fabric. The selection of a particular dyeing method depends on several factors including the form of textile material (fiber, yarn, knitted, or woven fabric), availability of suitable equipment in the mill, and batch size of the textile material.

7.12 Dyeing Machinery

The commonly used dyeing machines are as follows:

7.12.1 Exhaust Dyeing Machines

– *Package dyeing machine*: mainly used for dyeing yarn in package form
– *Winch or beck dyeing machine*: mainly used for dyeing knitted fabrics (in rope form) at atmospheric pressure but may also be used for woven fabrics
– *Jet dyeing machine*: mainly used for dyeing knitted fabrics (in rope form) at atmospheric or higher pressure but may also be used for woven fabrics
– *Jigger dyeing machine*: mainly used for dyeing woven fabrics (in open-width form)

7.12.2 Pad-Dyeing Machines

– *Pad-batch dyeing machine*: used for dyeing fabrics in the open-width form in a semicontinuous manner
– *Pad-steam dyeing range*: mainly used for dyeing cotton fabrics in the open-width form in a full-continuous manner
– *Pad-thermosol dyeing range*: mainly used for dyeing polyester and polyester/cotton-blended fabrics in the open-width form in a full-continuous manner
– *Stenter*: mainly used for simultaneous finishing and dyeing of fabrics with pigments

7.13 Blend Dyeing

Textile fabrics comprising a blend of more than one type of fiber can be dyed with suitable dyes to achieve different dyeing effects. In "union dyeing" both the fibers in a two-fiber blend (e.g., polyester/cotton) are dyed to have the same shade. In "cross-dyeing," each fiber component in a blend is dyed in a different shade. In "tone-on-tone dyeing" two fibers in a blend (e.g., cotton/viscose rayon) are dyed with the same class of dye but the two types of fibers have different depths of shade.

7.14 Textile Printing

The word "printing" is derived from the Latin word meaning "pressing" and implies the application of "pressure." Printing can be considered as "localized dyeing" and comprises the application of one or more dyes or pigments on textile materials in the form of a design or pattern [7]. Unlike dyeing, printing designs or patterns are usually printed on only one side of the fabric.

7.15 Common Styles of Textile Printing

7.15.1 Direct Printing

In direct printing, a color pattern is printed directly from a dye or pigment paste onto a textile substrate without any prior mordanting step or a follow-up step of dyeing.

7.15.2 Transfer Printing

In transfer printing, a design is printed first on a flexible nontextile substrate (e.g., paper) and later transferred from the paper to a textile substrate.

7.15.3 Discharge Printing

In discharge printing, a textile fabric is first dyed with a suitable dye and then the dye is selectively destroyed from certain areas of the fabric to look at a printed pattern.

7.15.4 Resist Printing

In resist printing, the fabric is first printed with a resist agent and then dyed. On dyeing, the fabric attains color only in areas where the resist agent is not present. After dyeing, the resist agent is removed and the fabric looks like a printed pattern.

7.16 Common Methods of Textile Printing

7.16.1 Block Printing

Block printing is an old method of printing that involves the use of wooden blocks with raised printing surfaces, which are inked and then pressed on to the fabric. This printing method is used only on a small scale or in the cottage industry and is not used at an industrial scale because of less flexibility and productivity.

7.16.2 Screen Printing

Screen printing is the most commonly used printing method on an industrial scale. There are two main types of screen printing: flat-bed screen printing and rotary screen printing. Flat-bed screen printing can be manual or automatic. Rotary screen printing is usually automatic and gives the highest printing productivity. Screen printing involves passing the print paste onto fabric through a mesh or screen which has some open and some blocked areas according to the desired print pattern. The print design obtained on the fabric depends on the pattern of the open areas of the screen.

7.16.3 Roller Printing

The roller printing is done by making use of heavy copper rollers engraved with a pattern. A separate roller is used for printing each color in the pattern. Due to low productivity, the roller printing method has been almost completely replaced by rotary screen printing.

7.16.4 Digital Printing

Digital ink-jet printing is one of the most modern ways of printing textile fabrics. This method can be used for most of the commercially available fabrics. In this method, a printing pattern can be directly printed from the computer onto the fabric with an ink-jet printer, without any need for making printing screens or engraved rollers. The design-to-print lead time is minimum in digital ink-jet printing and complex designs of photographic quality can be promptly printed. However, as compared to rotary screen printing, the productivity of ink-jet printing is very low. Hence, the method is mostly used for very short production runs or for printing smaller articles such as flags and banners [8, 9].

7.17 Main Steps in Direct Printing

7.17.1 Fabric Preparation

Good fabric preparation is necessary for obtaining good-quality printing results. Before printing, the fabric should be free from any impurities and protruding fibers. This is accomplished with appropriate singeing, desizing, scouring, and bleaching. Fabrics, especially containing thermoplastic fibers such as polyester, are made dimensionally stable by heat-setting before printing so that the print designs do not get changed by any subsequent shrinkage in the fabric.

7.17.2 Print-Paste Preparation

The print paste consists of a thickened solution or dispersion of dyes or pigments. The type of dye used depends on the fiber composition of the fabric to be printed. However, pigments can be used for printing fabrics made from all types of fibers. The use of a binder is essential in the case of pigment printing, which is not required in the case of printing with dyes. For making a printing paste of suitable viscosity, thickening agents or thickeners are used. The type of thickener used usually depends on the type of colorant used in printing. Viscosity is a very important parameter of the print paste because it determines the amount as well as the spread of the print paste applied during printing, ultimately affecting the penetration and sharpness of the prints. The stability of the print paste is also very important, which depends on the compatibility of the thickener with the colorant and auxiliaries and other factors such as the pH of the paste. Apart from the thickener, the use of any other auxiliary depends on the type of colorant used. Some examples of thickening agents include starch and its derivatives; British gum; locust bean gum; guar gum; gum tragacanth; gum Arabic; sodium alginate; and acrylate copolymers.

7.17.3 Printing

The application of the print paste can be accomplished by any suitable equipment and method that is available in a mill or lab, for example, flat-bed screen printing, rotary printing, or roller printing. The exact printing parameters depend on the type of process and equipment used.

7.17.4 Drying

After printing, the fabric is dried (usually by hot air) to prevent any accidental distortion or smearing of the print.

7.17.5 Fixation

For the fixation of the colorants, the printed and dried fabric is passed through a steamer or high-temperature hot-air curing/fixation/aging machine. The temperature and time of fixation depend on the type of colorant/auxiliary system used for printing.

7.17.6 Washing-Off

After fixation, the fabric is usually subjected to a washing-off treatment to remove any unfixed dye, thickener, or any other left-over auxiliaries from the fabric. The washing-off process is sometimes not necessary for fabrics printed with pigment/binder systems.

7.18 Textile Finishing

Finishing comprises final processes in the textile processing sequence to improve the appearance, hand-feel, or other aesthetics of the textiles or to add any extra functionality such as water repellency or flame retardancy. There are two broad categories of finishing:
– Chemical finishing
– Mechanical finishing

Commonly used chemical finishes include: softening, stiffening/hand-building, easy-care/wrinkle-recovery/durable-press, water/oil-repellent, soil-repellent, soil release, flame retardant, antislip, antistatic, antipilling, antimicrobial, elastomeric, UV protection, insect-resistant/moth protection, biopolishing, fragrance, moisture management, temperature adaptability, and finishes to improve color fastness of the dyed or printed fabrics [10]. Commonly used mechanical finishes include calendaring, compressive shrinkage/Sanforizing, raising, emerizing/sueding/peaching, and shearing/cropping [11].

Although both of the above categories of finishing are accomplished with the help of some machine, in chemical finishing the final effect obtained on the textiles is primarily due to the chemicals used in finishing. In mechanical finishing, the final effect obtained on the textiles is primarily due to some mechanical action on the fabric by the machine. Chemical finishing results in a change in the chemical composition of the fabric. Most of the chemical finishes do not result in a change in the fabric's appearance but may result in a change in some other physical and mechanical properties of the fabric. Mechanical finishing does not result in any change in the elemental composition of the fabric. However, most mechanical finishes alter the fabric's appearance.

In chemical finishing, relatively minor amounts of chemicals (often < 5 g/m^2 of the fabric) are applied on both sides of the fabric through padding or impregnation. In coating, relatively high levels of chemicals (15–50 g/m^2 of the fabric or even more) are applied on usually one side of the fabric (although sometimes fabrics may be coated on both sides).

From the durability perspective, there are three categories of finishes:
- Durable finishes undergo repeated laundering or dry cleaning without losing effectiveness. Durable finishes are usually expected to last more than 50 laundering cycles on the fabric.
- Nondurable finishes do not withstand washing or dry cleaning, and these are applied to textiles that are disposable or not supposed to be washed, or when the finishing effect is temporarily required.
- Semidurable finishes have their life in-between that of durable and nondurable finishes.

More than one type of finish can be combined in one process if they are compatible. The compatibility of finishes in a single formulation means that the finishes do not result in the formation of precipitates or instability of the formulation. For example, anionic finishes may not be compatible with cationic finishes. Combined finishes should also be compatible in terms of their final effect on the fabric. Some finishes may be contradictory. For example, hydrophobic finishes and hydrophilic finishes; stiffening finishes and softening finishes. Some finishes may result in more than one effect. For example, some fabric-softening finishes may also make the fabric hydrophobic. Some softeners may also have an antistatic effect on the fabric. Some finishes may also have an undesirable side effect. For example, the application of a crease recovery finish may result in a loss of fabric strength.

There are two primary requirements that chemical finishes should meet:
- Optimum desired effect in fabric properties at the lowest possible cost of the chemicals and the process;
- Possible adaptation of the finishing effects, according to the customers' requirements.

Some secondary requirements of chemical finishes include:
- Suitability of the finish for all fibers in all forms;
- Desired durability of the finishing effect;
- No loss of desirable fabric properties, for example, appearance, hand-feel, strength, comfort, and abrasion resistance;
- No yellowing of the white finished fabric;
- No change in the color of the dyed or printed fabric;
- Safe to use and simple to apply on the fabric;
- Good storage stability and shelf life;
- Good compatibility with other ingredients of the finishing formulation;
- Easy correction in case of faulty finish application;
- Sustainability and no harmful impact on the environment.

Factors that are commonly considered for proper formulation of the chemical finishes include:
–   The type of textile being treated (fiber type, yarn type, fabric construction);
–   The performance requirement of the finishes (the extent of desired effect and its durability);
–   The cost-to-benefit ratio;
–   Limitations imposed by the available machinery or environmental issues;
–   Compatibility of various components of the formulation.

7.19 Chemical Finishing

7.19.1 Softening

Softening is one of the most commonly used textile finishes. Fabric softness usually depends upon four measurable fabric characteristics, that is, coefficient of friction, flexibility/bendability, compressibility, and elasticity. Objectives for the application of chemical softeners include improvement in hand-feel, drape, tear resistance, or sewability of the fabric. Softeners lubricate the fibers, decrease the coefficient of friction, improve fabric smoothness, and may also lower the glass transition temperature of the polymer. The lubricating effect of the softeners improves fabric sewability by reducing friction between the sewing needle and the yarns in the fabric. Higher friction between the sewing needle and the fabric may cause a rise in the needle temperature, leading to possible needle breakage and yarn damage. Softeners increase fabric softness by reducing the fabric coefficient of friction or by internal lubrication or plasticization of the fibers and yarns or by reducing the glass transition temperature of the polymer.

Based on their ionic nature, softeners may be classified as nonionic, anionic, cationic, or amphoteric. Nonionic softeners do not have any electrical charge and may be based on hydrocarbons, ethylene oxides, or silicones. Because of having no electrical charge, they have good compatibility with other finishes. Some nonionic softeners may also be effective in improving fabric wrinkle resistance, abrasion resistance, tear strength, and sewability. They have a low tendency to make yellow the white fabric or change the color of the dyed fabrics. Some silicone-based softeners may be expensive and may also be hydrophobic, thus being unsuitable for softening towel fabrics.

Anionic softeners have a negative charge and may be based on fatty esters, waxes, or sulfonated oils. They often retain good rewetting properties for the treated fabric. However, due to no inherent attraction for the cellulosic materials, they may not exhaust from the aqueous bath onto the textiles. Moreover, they have limited durability in washing and dry cleaning.

Cationic softeners are positively charged and may be based on quaternary ammonium salts, fatty amines, imidazolines, or aminoamides. Due to their affinity for cellulosic materials, they can easily exhaust from an aqueous bath onto the textile material. Cationic softeners often give good softness with low add-on and improve fabric abrasion and tear resistance of the fabric. However, they may be incompatible with anionic finishes and auxiliaries and may make the fabric hydrophobic. They also have the potential to yellow the white fabrics and result in shade change of the dyed fabrics.

Amphoteric softeners may have a positive or negative charge depending on the pH of the application bath. However, these softeners are less commonly used as compared to other classes of softeners.

In addition to subjective assessment by hand, the softness of the fabric treated with softeners can be assessed using a fabric touch tester, softometer, handl-o-meter, Kawabata evaluation system, fabric stiffness tester, and bending length tester or Fabric Assurance by Simple Testing system.

7.19.2 Stiffening/Hand-Building

Hand-building finishes are applied to the fabric to increase: fabric stiffness, stability, bulkiness, or weight. Objectives of improving fabric stiffness include: improving fabric handling during cutting and sewing, stabilizing a limp fabric, or improving fabric appearance.

Hand-building finishes may be nondurable or durable. Nondurable finishes include starch, PVA, and CMC. Durable hand-building finishes may be thermoplastic or thermosetting. Thermoplastic hand-builders include vinyl polymers and examples of thermosetting hand-builders include melamine formaldehyde and urea formaldehyde. The choice of hand-building finishes depends on cost, ease of application, degree of stiffness, bulkiness, stability, and durability required. The assessment of the fabrics treated with hand-building finishes may be done by determining the increase in fabric weight per unit area, bending length, stiffness, or flexural rigidity.

7.19.3 Easy-Care, Wrinkle Recovery, and Durable Press Finishing

The cellulose antiswelling or cellulose cross-linking finishes may be applied on cellulosic fabrics (e.g., cotton) to achieve different effects such as a smooth wrinkle-free appearance after washing without ironing (easy care and wrinkle resistance), retention of intentional creases after washing (durable press), and shrink resistance after washing (dimensional stability). Such finishes may also be used to increase the fabric's pilling resistance and pile resilience or to enhance the durability of dyes, pigments, or other finishes. These finishes may result in a reduction in fabric elasticity, flexibility, abrasion resistance, and tear strength.

The woven fabrics with tighter fabric density containing yarns with higher twist levels and coarser fibers and higher hydrophilic character are more prone to wrinkling and vice versa. Woven fabrics made from cotton tend to wrinkle more as compared to polyester or blended fabrics. Cotton fibers can absorb moisture, which can disrupt hydrogen bonding between hydroxyl groups of the cellulose polymer chains, facilitating the chain movement and formation of new hydrogen bonds in a new wrinkle configuration. The process can be reversed by steam-ironing the fabric. The application of cellulosic crosslinking finishes results in more permanent cross-links between the cellulose chains, restricting their free movement on moisture absorption, thus limiting their tendency to wrinkling.

Different types of easy-care/wrinkle recovery finishes include urea/formaldehyde, melamine formaldehyde, dimethylol dihydroxy ethylene urea (DMDHEU), methylated/modified DMDHEU, and polycarboxylic acids such as butane tetracarboxylic acid. The most commonly used easy-care/wrinkle-recovery finish is DMDHEU, which is usually applied in the presence of a catalyst such as magnesium chloride.

In precure finishing, the fabric is cured immediately after drying followed by the impregnation in the finishing liquor, that is, before garment manufacturing. The process results in improved wrinkle recovery of the fabric. In postcure finishing, the fabric is impregnated in the finishing liquor, dried, and then converted into garment form before the curing step. The postcure process is usually used for durable press effects such as crease retention in trousers or pleated skirts. In addition to precure and postcure methods, the crosslinking finish may also be applied after garment manufacturing, for example, spraying a denim garment with the finishing liquor for obtaining creases at specific places on the garment.

Different methods are used for fabric assessment after wrinkle recovery or durable press finishing. These methods include:
- AATCC Test Method 88C: Retention of creases in fabrics after repeated home laundering
- AATCC Test Method 124: Smoothness appearance of fabrics after repeated home laundering
- ISO 2313: Determination of the recovery from creasing of a horizontally folded specimen of fabric by measuring the angle of recovery

7.19.4 Water-Repellent Finishing

Water-repellent fabrics can resist wetting whereas water-proof fabrics are impermeable to water even at high hydrostatic pressure and also usually impervious to air. The water repellency of a fabric depends on several factors including the nature of the fibers, yarn structure, fabric porosity, finished applied, and water impact force. Some fabrics such as those made from cotton easily wet out as compared to those made from hydrophobic fibers such as polypropylene. Generally, fabrics or other surfaces which have high surface-free energy have better wetting as compared to those of lower surface-free energy.

Different chemicals may be applied on the fabric to lower the surface-free energy of fabrics than water surface tension to decrease their wetting ability and increase their water repellency. Three main types of water-repellent chemicals are wax-based repellents, silicone-based repellents, and fluorocarbon-based repellents. Wax-based repellents are usually the cheapest while fluorocarbon-based repellents are usually the most expensive and the most durable. While wax and silicone-based chemicals may result in water repellency only, fluorocarbons result in water as well as oil repellency in the fabric.

The water contact angle is a good indication of the water repellency of a fabric. The higher the contact angle, the higher will be the water repellency of the fabric. Fabrics with a water contact angle of greater than 90 may be considered water-repellent while fabric with a contact angle greater than 130–150 may be considered super-repellent fabrics. Different standard test methods can be used to evaluate the water repellency of fabrics. One such method is AATCC Test Method 22–2001: water repellency-spray test. This method involves spraying water against the taut fabric surface under controlled conditions. The degree of wetting is rated from 0 to 5 scale, where 0 refers to complete wetting and 5 to no wetting.

7.19.5 Stain Resistance Finishing

Staining refers to localized soiling of a textile material by oil, grease, dry particulate matter, oil-borne stains, or water-borne stains, whereas soiling is the overall contamination or discoloration of a textile material. An example of staining may be a drop of tea, oil, or ketchup on a shirt. Stain repellency is the ability of a fabric to resist penetration by liquid soils under static conditions, that is, conditions under which the liquid is not forced into the fabric by external influence other than the weight of the liquid drop and capillary forces. Stain resistance is the degree to which a fabric, stained under dynamic conditions, can be returned to its unstained state by wiping or blotting the fabric surface.

Stain resistance finishing is usually obtained by the application of silicones or fluorocarbons which lower the surface-free energy of fabrics and make them water and oil-repellent. Silicones may result in resistance to only water-borne stains but fluorocarbons give resistance to both water-borne and oily stains.

7.19.6 Stain or Soil Release Finishing

Stain or soil release is the ability of a fabric to be cleaned easily by laundering. Stain-release properties are important for those textiles that can be washed whereas stain-resistant properties are important for upholstery, carpets, or such other fabrics that cannot be conveniently laundered. Soil release, particularly of oily stains, is usually difficult in textiles made from hydrophobic fibers such as polyester. Soil release properties may be imparted by applying hydrophilic treatments to hydrophobic textiles. For example, low–molecular-weight block copolymers of hydrophilic segments like polyoxyethylene can be used to improve the soil release properties of polyester fabrics. Conventional soil resistance finishes deteriorate soil release properties of fabrics but dual action fluorocarbons comprising a block copolymer of fluorocarbon component and a hydrophilic polyoxyethylene component have good soil resistance as well as good soil release properties.

7.19.7 Flame-Retardant Finishing

The flammability of textiles pertains to their relative ease of ignition and relative ability to sustain combustion. Fabric can be considered flame-resistant if it does not burn or does not continue to burn when subjected to a flame or heat source, with or without removal of the source. A chemical applied to a fabric to impart flame resistance is called a flame-retardant. Different factors affecting the flammability of textiles include the type of fiber, yarn structure, fabric structure, and any chemicals/coatings applied to the fabric. Three necessary components for a fire are fuel, heat, and oxygen. Flame-retardant finishes improve flame resistance by masking or removing any one or more components that are required for burning. Some flame retardants promote the char formation of cellulosic textiles, thus removing the fuel in the form of fiber which is required to sustain fire. Some flame retardants form an insulating layer on the fibers, thus restricting their access to air which is required to sustain burning.

Some chemical flame retardants may be durable even after more than 50 laundering cycles while others may be nondurable and washed away after single laundering. Some examples of nondurable flame retardants for cellulosic textiles include ammonium phosphate, di-ammonium phosphate, ammonium chloride, ammonium bromide, and borax-boric acid mixtures. Examples of durable flame retardants for cellulosic textiles include tetrakishydroxyphosphonium chloride/ammonia and N-methylol dimethylphosphonopropionamide (Pyrovatex). Most flame retardants are applied by the pad-dry-cure method, although some may be applied using the exhaust method.

Two commonly used methods for assessing the flammability of textiles are the vertical flame test and the 45 C test, where the fabric test specimens are held vertically or at 45 C, respectively. Determination of limiting oxygen index (LOI) is also another useful method for characterizing fabrics treated with flame retardants. LOI measures the amount of oxygen required to sustain burning. The LOI of untreated cotton is around 18.5. The higher the LOI, the lower the flammability of the textile material.

7.19.8 Antimicrobial Finishing

The antimicrobial finishes are those which work against the microbes (e.g., bacteria, fungi, mildew, and virus) by inhibiting their growth or by killing them altogether.

Biostatic finishes inhibit the growth of microbes whereas biocidal finishes kill the microbes. Microbes may not only result in the loss of some functional properties of textiles such as loss in strength but may also result in bad odor, staining, and discoloration. Their undesirable effects on humans may include perspiration smell, eczema, irritation, allergy, or even infection or disease.

The two main types of antimicrobial finishes are the leaching type and the chemically bound type. Leaching-type antimicrobial finishes leave the textile and chemically enter or react with the microorganism through a controlled release mechanism. They are effective on the fabric surface and in small surrounding environments but eventually, the reservoir is depleted and the finish may be no longer effective. Chemically bound-type antimicrobials may be more durable, and they do not leave the textiles and control only those microbes which come into contact with the textile. Mechanisms of various antimicrobial finishes include: preventing microbial cell reproduction, blocking enzymes, reacting with the cell membrane, destruction of cell walls, or poisoning the cells from within.

Some examples of antimicrobial finishes include triclosan, quaternary ammonium compounds (e.g., octadecylaminodimethyltrimethoxysilylpropylammonium chloride), polyhexamethylene biguanide, chitosan, and silver nanoparticles. Textiles treated with antimicrobial finishes may be characterized by using qualitative as well as quantitative methods. In quantitative methods, the number of bacteria or bacterial colonies is counted, after an opportune contact time. The qualitative tests indicate the presence or absence of antibacterial activity (e.g., through a zone of inhibition).

7.19.9 Biopolishing

Biopolishing is a process of treatment of cotton and other cellulosic textiles with cellulases enzymes to remove protruding fibers from textile fabrics and produce a softer and smoother hand-feel. Biopolishing also reduces fabric pilling tendency and improves fabric luster, brightness, and drape. Biopolishing can also be used as a replacement or as a supplementary process for denim stone washing. In biopolishing, the celluase enzymes are used to break down the cellulose polymer chains of the surface fibers by hydrolysis, which are then easily removed during washing due to increased water solubility. Important parameters of cellulases biopolishing process include the concentration of the enzyme, pH, temperature, any surfactant used, process time, and agitation/mechanical action during the process.

7.19.10 Moisture Management Finishing

Moisture management refers to the engineered movement of perspiration from a garment's near-to-skin side through the fabric to the outer side for an evaporative cooling effect for the wearer. Garments made from hydrophilic fibers such as cotton are good in perspiration absorption but poor in wicking and drying because of hydrogen bond formation with water. On the other hand, garments made from hydrophobic fibers such as polyester are poor in sweat absorption but good in wicking and drying due to a lack of hydrogen-bonding sites. The controlled application of some hydrophilic finishes on fabrics made from hydrophobic fibers has been reported to improve the moisture management properties of such fabrics.

7.19.11 Antistatic Finishing

The garments made purely from hydrophobic fibers such as polyester tend to develop a static charge, resulting in clinging of garments to the wearer's body and/or an annoying crackling sound while wearing on or taking off a garment. The tendency to accumulate static charge can be decreased by increasing the fabric conductivity and/or reducing the frictional forces by applying suitable lubricating agents. The hygroscopic finishes can be used to increase fabric conductivity. Some examples of nondurable antistatic finishes include polyethylene glycol and polyethylene oxide compounds. Some polyamines may be reacted with polyglycols for the durable hydrophilic finishing of textiles. Deposition of carbon or metallic (e.g., nano-silver) coatings may also result in increased fabric conductivity and reduced static charge accumulation.

7.19.12 Optical Brightening

Fabrics that are to be finished white may be treated with optical brightening agents or fluorescent brightening agents to increase their brightness. Such agents may be considered colorless dyes, which when present on the fabric can absorb the wavelengths of light in the UV region and reflect them in the visible region, thus making the fabric appear brighter as compared to the untreated fabric.

7.20 Mechanical Finishing

7.20.1 Napping

Napping is a process of raising protruding fibers from the surface yarns of a fabric to form a raised pile. The objective of raising protruding fibers at the fabric surface is to improve the fabric's hand-feel and the ability to give warmth. The warmth of the fabric is increased due to its increased ability to entrap air because of the raised fibers.

Air, being a good heat insulator, improves the fabric's heat retention ability. Examples of napped fabrics include fleece and flannel.

Napping or raising of fibers at the fabric surface is achieved by passing the fabric across rapidly rotating wire-covered rollers. The wire-covered rollers of a napping machine usually rotate against the fabric's direction of passage. The extremely sharp wires on the rollers pluck and raise fabric surface fibers. The main parameters of the napping machine include machine type, speed of the rollers, condition of roller wires, yarn and fabric construction, fabric speed and tension, and presence of any finishing agent (e.g., softener) on the fabric.

7.20.2 Shearing

Shearing is a process of removing fuzz or fibers protruding on a fabric surface to produce an even surface with a uniform height of the pile of raised fibers on the fabric. Shearing is usually done after the napping process to cut the raised fibers for producing an even pile height on the fabric surface. Shearing may also be done for fabrics that are not napped, and the process may be used as an alternative to singeing and biopolishing for removing the protruding fibers from a fabric surface. The shearing process results in a smoother fabric surface and improvement in the pilling resistance of the fabric. Shearing is achieved by passing the fabric close to a revolving spiral blade in contact with a ledger plate. The spiral place and the ledge plate together resemble the action of the blades of a pair of scissors, cutting the fibers on the fabric surface.

7.20.3 Sueding/Emerizing

Sueding or emerizing is a process of raising very short fibers on the fabric surface by passing the fabric over sandpaper-covered/emery rolls. The purpose of sueding or emerizing is to produce a special hand-feel on the fabric that is similar to suede leather or the skin of peach fruit. The length of raised fibers after sueding/emerizing is usually shorter as compared to that obtained after napping. Important parameters of the sueding process include the speed of the sand rolls, coarseness of the sandpaper, yarn and fabric construction, fabric speed, the pressure between the fabric and the sandpaper, direction of rotation of rolls to the direction of fabric passage through the machine, and the type of sueding machine. Two main types of sueding machines are single cylinder and multicylinder.

7.20.4 Calendering

Calendering is a process of imparting luster and smoothness to a fabric by passing it between pressurized rollers. If the moist fabric is passed through the pressurized rollers, the calendered fabric will be quite similar to steam-ironed fabric. The main effect produced by calendering includes the reduction in fabric thickness, compaction of weave structure, change in fabric handle, and change in fabric luster. Variation in any of the following parameters results in different effects produced by calendering: fiber content, fabric construction, moisture in the fabric before calendering, any chemical finish applied before calendering, the temperature of calender roll(s), relative speeds of fabric and rolls, roll composition and configuration, pressure applied, and the number of times the fabric is passed through the calender rolls. One of the calender rolls is usually made of stainless steel (which may be heated to the required temperature) while the other may be covered with highly compressed cotton, paper, or synthetic material. Common types of calenders used in the industry include three-roll friction calenders, Schreiner calendars, and embossing calenders.

7.20.5 Compacting

Compacting is a process of mechanical compression of fabrics in the lengthwise direction to minimize their tendency to shrink during consumer use. The compacting process has little effect on the widthwise shrinkage of the fabric. The compacting process also results in an increase in fabric areal density and thickness and a reduction in the overall fabric length yardage. Different types of machines can be used for compacting including the trademark Sanforizer (also known as blanket compacter), the heated roll and shoe compactor, and the blade compactor. The compacting process was originally devised for woven fabrics but the process was also adapted later for knitted fabrics.

7.20.6 Relaxation Drying

Relaxation drying is a process of drying knitted fabrics by passing them through a dryer with overfeeding and minimum tension so that the fabric may be able to shrink during fabric processing rather than later in garment form during consumer use. Conveyer-belt-type relaxation dryers are the most popular, although several other designs also exist. Important factors in relaxation drying include the amount of overfeeding, the spreading of tubular knits before entering the dryer, the moisture content in the incoming fabric, the mechanical action of the dryer, the temperature of the dryer, and the presence of any finish on the fabric before the process.

7.20.7 Decatizing

Decatizing is a process for improving the luster and smoothness of wool fabrics by layering the wool fabric between heavy cotton fabrics on a roller and exposing it to steam. The pressure and steam of the process flatten the wool fabric, improving its luster and smoothness.

7.21 Developments in Sustainable Textile Processing

In the entire textile value chain, textile processing is considered by far the most intensive step in terms of water, chemicals, and energy consumption. According to estimates, around 280,000 tons of dyes are discharged every year as industrial effluent, with an average of 150 L of water consumed for dyeing every 1 kg of textile material [11]. In the wake of increasing demand for more sustainable textile processing, several innovations have been reported in recent years, which can be broadly classified into two groups: sustainable chemistry and sustainable application technology or machinery.

7.21.1 Sustainable Chemistry

Only a few commercially viable new developments have been reported for sustainable pretreatment of textiles during the last decade. Enzymes have traditionally enjoyed widespread application in desizing. A significant segment of the industry has also started using enzymes for the scouring of cotton fabrics as a replacement for the alkaline process using caustic soda. The bioscouring process using enzymes allows savings in energy, time, and use of harsh chemicals, along with improving worker safety. However, the cotton bleaching process has not so far seen a widespread application of enzymes.

Cationic pretreatment of cotton has also been explored by some industries for making the cotton cationic and more receptive to the dye for increased dye uptake and reduced dye discharge in the effluent. The process also allows a reduction in the use of salt during dyeing.

There has been a renewed interest in the extraction and application of natural dyes, as an alternative to synthetic dyes. For example, plant-based indigo is becoming a highly desirable colorant for denim, as a replacement for synthetic indigo dye based on petrochemicals [12].

Carbon-negative colorants from algae have been introduced. One such example is a black pigment produced from algae waste biomass [13]. The product has been successfully used in small prints. However, it has yet to be explored for large-scale printing in textiles. Algadye 3.0 is an algae-based dye formulation that has been claimed to be suitable to dye all types of synthetic, natural, and protein-based fibers [14]. However, limited reports are available on the fastness properties of this colorant on different textile materials.

High-performing black pigments from wood waste have been introduced by Nature Coatings [15]. Produced through a closed-loop manufacturing process with minimum emission of CO_2 and greenhouse gases, the pigments are claimed to be suitable for rotary and flat-screen printing.

Another recent development is the microbial pigments produced from naturally occurring and artificially grown genetically modified microbes. Examples of such products include Colorfix [16], Huue [17], Pili [18], and KBCols Sciences [19].

An interesting innovation is the use of captured carbon as a feedstock for producing pigment not only shifting away from using nonrenewable petrochemical-based pigments but also decreasing the impact of greenhouse gases. The black pigment produced by Graviky [20] is claimed to be suitable for dyeing polyester and paper.

Recycling dyes is another trend for improving the sustainability of the textile dyeing process. Recycling of dyes can be accomplished either by chemically or biologically recovering dyes from textile waste or converting the textile waste into very fine crystallized powder and dyeing new textiles with it. It has been claimed that the powdered textile waste dye can be used as a pigment suspension for application through various dyeing and printing methods [21].

In textile finishing, significant efforts are being made for finding durable water and oil-repellent finishes which are free from perfluorochemicals (PFCs). PFCs have been reported to be harmful and toxic for humans as well as for the environment, leading to efforts for finding alternative bio-based solutions or chemistries including silicone. Perfluoroalkyl (PFOAs) substances have been linked with negative human thyroid function as well as testicular and kidney cancer. Similarly, perfluorooctanesulfonic acid has been linked with infertility and adverse developmental effects.

Recently high performing durable water and oil-repellent finishes have become available, which are free from PFOAs and can be applied on cotton and synthetic fibers using traditional methods.

Although several new textile processing chemistries have surfaced in recent years, they are still facing fierce competition with traditional chemistries in terms of durability, scalability, reproducibility, and cost.

7.21.2 Sustainable Application Technology or Machinery

Plasma is one of the most promising water-less technologies introduced recently in textile processing. Formed by ionizing a gas, plasma is the fourth state of matter that can be employed for removing fabric impurities, activating its surface as well as depositing a coating. The technology does not result in any effluent and has lower energy consumption as compared to water-based processing technologies.

A newly introduced multiplexed laser surface enhancement system can create a quantum mechanical energy milieu using laser plasma. This technology can be used as a continuous process for the pretreatment and finishing of all types of fabrics. Proprietary atmospheric plasma technologies are also available to replace the conventional bleaching process.

Recently, spray-dyeing systems have made it possible to apply the exact amount of colorant and chemical finishes directly on the fabric through a highly efficient digitally controlled process. Unlike digital printing, the spray-dyeing system can produce deeply penetrated solid colors on the fabric using traditional dyes instead of special inks used in digital printing.

One such a system is developed by Alchemie™ for both dyeing and finishing, along with a proprietary fixation process using infrared radiation. The technology is claimed to be suitable for polyester and cotton fabrics using traditional dyes and finishes. Similarly, Imogo™ has also introduced a digital spray dyeing and finishing system using capillary forces to get deep absorption and penetration of the dyes or finishes.

Supercritical CO_2 dyeing has enabled the replacement of water as a solvent with supercritical CO_2 having flow properties like gas and solubility like a liquid. The close-loop waterless dyeing process results in zero-water effluent, using fewer chemicals and energy as compared to the conventional dyeing process.

Another innovation in sustainable textile processing is the advent of ultrasonic dyeing and finishing processes. Comprising thousands of microscopic bubbles, a large amount of energy is released when the bubbles burst. This energy is used to inject the desired chemistries into the fabric using less water and chemicals as compared to traditional processes. One such process has been developed by Sonovia™ for applying durable water-repellent and antimicrobial finishes. Another example is a process developed by Indidye™ for the extraction and application of natural dyes on cellulosic fibers.

Foam dyeing also offers a low-water coloration process using foam bubbles to carry the colorant and deposit it into the textile material when the bubble bursts. A proprietary process IndigoZERO™ has been developed to apply indigo dye to cotton yarns using foam technology with minimal use of water.

Dope dyeing also offers an alternative to conventional coloration whereby the colorant is added to the polymer dope before extruding the manmade filament yarns. The process does not apply to natural fibers but can only be used for man-made fibers.

High-quality textile printing is now possible through the inkjet-based digital printing method, steadily replacing traditional rotary screen printing, especially for shorter batches. A digitally controlled process uses a significantly lower amount of energy and no water at all for making the printing pastes as required for conventional screen printing. Using ink-filled depressed cells, digitally controlled Gravure printing can also be used as another almost waterless process with less energy consumption.

Ozone, a strong oxidizing agent, has been used in the treatment of jeans for fading to have a worn-out look with less use of chemicals, water, and energy. The close-loop technology allows the conversion of the remaining ozone back into oxygen at the end of the process. Laser is another fading technology used in denim finishing as a replacement for bleaching, sandblasting, or manual scraping.

Despite the availability of several sustainable alternatives to conventional water-intensive textile processes, the transition has so far been slow on a large scale. One possible reason is the high initial investment and limited ability to meet the performance standards in comparison with traditional technologies. With continuous improvement in innovative sustainable technologies, it is hoped that the industry will ultimately shift toward the best available technologies as far as sustainable textile processing is concerned.

References

[1] R. Shamey, and T. Hussein, "Critical solutions in the dyeing of cotton textile materials," Textile Progress, vol. 37, no. 1–2, pp. 1–84, 2005.

[2] Osthoff Singeing Technology, (accessed February 17, 2015 at http://www.osthoff-senge.com/en/tech nologie.html).

[3] C. Tomasino, Chemistry and technology of fabric preparation and finishing, NCSU, North Carolina, (accessed February 10, 2015 at http://infohouse.p2ric.org/ref/06/05815.pdf).

[4] V. B. Gupta, "Heat setting," Journal of Applied Polymer Science, vol. 83, pp. 586–609, 2002.

[5] M. Clark, *Handbook of Industrial Dyeing – Volume 1: Principles, Processes and Types of Dyes*. Cambridge, UK: Woodhead Publishing, 2011.

[6] M. Clark, *Handbook of Industrial Dyeing – Volume 2: Application of Dyes*. Cambridge, UK: Woodhead Publishing, 2011.

[7] W. C. Miles, *Textile Printing*, 2nd ed. Bradford, UK: Society of Dyers and Colorists, 2003.

[8] H. Ujiie, *Digital Printing of Textiles*. Cambridge, UK: Woodhead Publishing, 2006.

[9] C. CIE, *Ink Jet Textile Printing*. Cambridge, UK: Woodhead Publishing, 2015.

[10] W. D. Schindler, and P. J. Hauser, *Chemical Finishing of Textiles*. Cambridge, UK: Woodhead Publishing, 2004.

[11] T. Hussain, and A. Wahab, "A critical review of the current water conservation practices in textile wet processing," Journal of Cleaner Production, vol. 198, pp. 806–819, 2018.

[12] 00% plant-based indigo [Internet]. [cited 2022 Aug 3]. Available from: https://www.dyerecycle.com/.

[13] Carbon-back pigment from algae biomass [Internet]. [cited 2022 Aug 3]. Available from: https://livingink.co/.

[14] Algaeing [Internet]. [cited 2022 Aug 3]. Available from: https://algaeing.co/.

[15] High performing black pigments from wood waste [Internet]. [cited 2022 Aug 3]. Available from: https://naturecoatingsinc.com/.

[16] Colorfix: the science of sustainable color [Internet]. [cited 2022 Aug 3]. Available from: https://colorifix.com/colorifix-solutions/.

[17] Biodynthetic indigo [Internet]. [cited 2022 Aug 3]. Available from: https://www.huue.bio/.

[18] Pigments from microbial enzymes [Internet]. [cited 2022 Aug 3]. Available from: https://www.pili.bio/9/technology.

[19]   KBCols microbial pigments [Internet]. [cited 2022 Aug 3]. Available from: https://kbcolssciences. com.

[20]   Black pigments made from captured carbon [Internet]. [cited 2022 Aug 3]. Available from: https://www.graviky.com/.

[21]   Recycron pigment powders from waste textiles [Internet]. [cited 2022 Aug 4]. Available from: https://recycrom.com/tech-explained/.

Abher Rasheed

8 Clothing

Abstract: Garment manufacturing is a complex process that involves multiple stages, from designing to production. This chapter will provide an overview of the various stages of garment manufacturing, including pattern making, cutting, sewing, and finishing. It will explore the role of technology in the manufacturing process, including computer-aided design (CAD), spreading and automated cutting machines. The chapter will also examine the structure of global apparel industry, including the stakeholders involved in this business. The chapter will illustrate the classification of garments. Ultimately, this chapter will argue that garment manufacturing is a critical component of the clothing industry and requires a holistic approach that balances creativity, efficiency, and ethics to create high-quality garments that meet the needs of both consumers and the planet.

Keywords: Clothing, Garment, Apparel, manufacturing

8.1 Introduction

Before going into the details of the clothing manufacturing process, the readers need to know a few basic definitions.

8.2 What Is Garment/Apparel/ Clothing?

"Any object which can be used to wear is termed as garment/apparel/clothing" (e.g., pants, shirt, and sweater). Another definition of the same terminology is "the 3-dimensional shape of cloth which is used to cover the body." They can also be used to achieve a specific function, for example, fire resistance or cut resistance. Garments are a part of apparel, whereas apparel is a broad terminology. For instance, shoes and hats are also included in apparel.

8.3 Types of Garments

Garments can be classified based on their use, shape, or fashion. Garments are classified into three types based on their use:

https://doi.org/10.1515/9783110799415-008

- Tops
- Bottoms
- Undergarments

Tops: The garments which cover the trunk portion of the body are known as tops (e.g., shirts, jackets, and coats).

Bottoms: The garments which cover the lower part of the body are called bottoms (e.g., trousers, pants, and shorts).

Undergarments: Any garments worn under the visible outer clothes, usually next to the skin, are called undergarments. Vests, underwear, briefs, and brassieres are examples of undergarments. These garments are used for different purposes and on different parts of the body. Underwear and brassieres are worn to provide support to specific body parts. Vests are used in warm seasons to absorb perspiration so that the transfer of sweat to the outer clothing can be minimized.

Garments are classified into two types on the basis of silhouette/shape:
- Fitted garment
- Loose garment

Fitted garment: The garments which fit closely to the body and depict the shape of the body are called fitted garments. Undergarments are usually designed to be fitted to the body. Leggings, yoga dresses, and formal dresses are other examples of fitted garments.

Loose garment: These garments are manufactured with an extra amount of ease allowance and the garment does not stick to the body tightly. Evening wears are an example of loose garments.

Based on fashion, garments are classified into four types (Figure 8.1):
- Staple products
- Semistyled products
- Styled products
- Fashioned products

Staple products: These garments are manufactured without any change for a long time. With slight alterations, these garments are manufactured as such for many years. Sometimes the fabric is slightly altered. These products have more lead time but very low-profit margins. Examples: vests, men's briefs, and workwear.

Semistyled products: These are the basic type of garments which bear slight changes in style. These types of garments have more alterations than staple products and these are also manufactured as such for a long time. The alterations include color change, change in shape and patterns of pockets, cuffs, and collars. Example: men's formal shirts and trousers.

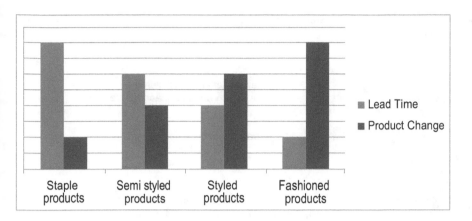

Fig. 8.1: Comparison of garment types based on lead time and product change.

Styled products: The garments which are altered quite often are known as styled products. These garments have more and quick alterations than semistyled products. For instance, change in length of the garment, change in the shape of the garment, change in cuffs and collars, and so on are some alterations. Examples: jackets and ladies' dresses.

Fashioned products: These garments are manufactured in different styles at a time. These garments are quickly altered and have many alterations. It is quite difficult to manufacture such products. It requires highly skilled labor and management. These products have minimum lead time and maximum profit [1]. Example: fashioned articles.

8.4 Global Apparel Industry

During the last decade, the apparel industry is shifting toward Asia. Earlier this industry flourished in the USA, the UK, and Europe. The most important reason for shifting this industry to Asia is the availability of relatively cheap labor. The other reason is that the developed countries are moving toward the industries which demand higher skills. For example, weapons, mechatronics, and telecommunications.

According to the World Trade Organization [2], China, Hong Kong, Bangladesh, India, and Turkey are the biggest clothing exporters in the world. Pakistan is the tenth biggest clothing exporter in the world.

The clothing exports of Bangladesh are four times higher than the clothing exports of Pakistan. Even though Bangladesh has to import all of its raw materials Pakistan is the fifth biggest cotton producer in the world. Pakistan has a very well-structured spinning and weaving industry. Pakistan can increase its clothing exports by using raw materials as its strength. In addition to that, improvement in quality may also impact positively the exports of Pakistan. The global apparel industry can be divided into four parts as shown in Fig. 8.2.

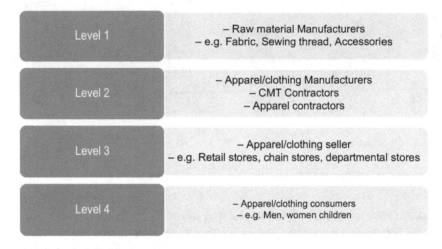

Fig. 8.2: Levels of the global apparel industry.

According to Fig. 8.2, the apparel industry is divided into four parts. We will discuss each level one by one.

Raw material manufacturers come at **level 1**. Raw material means the materials required for apparel/clothing such as fabric, sewing thread, labels, and packing material. In this respect spinning, weaving, knitting, and textile processing are working on level 1. The garment industry is acquiring its raw materials from these manufacturers.

Apparel manufacturers are at **level 2**. Apparel manufacturers are the organizations making finished products. Such companies normally have an organizational structure which has merchandizing/marketing, finance, operations, and production departments. Most of the apparel manufacturers in Pakistan are working with contractors while a few hire permanent labor. Operations like sewing, cutting, and finishing which require more labor are assigned to the contractors. The contractor is responsible for providing operators for each task and completing each consignment on time.

Sometimes apparel manufacturers have more orders than their operational capacity. In such cases, cut, manufacture, and trim (CMT) is an option which can be used. CMT is also a type of contract. In this contract, a manufacturer who has more orders than his capacity gives a contract to another manufacturer who is known as a CMT contractor. CMT contractor is responsible for all operations including cutting, stitching, and trimming. Some small-sized manufacturers who do not have their marketing/merchandizing setup only work as CMT contractors. Therefore, apparel manufacturers, CMT contractors, and labor contractors are working at level 2.

Apparel sellers are at **level 3**. Once the finished products are manufactured, they are sent to the sellers. Different types of sellers are as follows:

- Retail stores
- Chain stores
- Discount stores
- Departmental stores
- Mass merchandizers

Consumers are part of this system at **level 4**. It includes all men, women, and children in the world.

8.5 Structure of an Apparel Firm

There is a lot of variation in the structure of an apparel firm worldwide. Each firm creates its own structure keeping in view its resources and constraints. Figure 8.3 depicts the basic structure of an apparel industry.

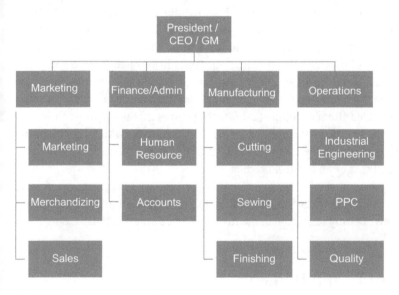

Fig. 8.3: Structure of an apparel firm.

8.6 Marketing Division

The basic responsibility of marketing is to satisfy the customers. They are also responsible to find out new customers. They contact different potential customers in the world and try to convince them to do business with their company. There are three basic functions of marketing:

The marketing section is responsible for scheduling the marketing calendar, product pricing, product planning, and customer relationships. In the majority of the organizations in Pakistan's apparel industry either there is no marketing department or it is not efficiently functioning. The reason is that they do not design the products. Normally, the production takes place according to the customer's requirements and instructions.

The basic objective of *merchandizing* is to provide the right product at the right place, at the right time, and at the right price. Therefore, there are four basic functions of merchandizing. The merchandizing department performs these functions by acting as a bridge between the customer and manufacturing. They continuously monitor the production of a certain order and in case of any problem, they correspond with the customer.

A merchandizer should be technically very strong because he/she has to deal with the whole manufacturing process. Unfortunately, in Pakistan merchandizing is full of people without much technical background. That is why the industry here loses many orders. For example, most of the merchandizers do not know the capacity of their units. Therefore, they accept orders which cannot be manufactured in their unit. In such cases, new machinery has to be purchased and the manufacturing cost is increased. In addition to that, machine layouts have to be changed again and again to make different types of products. Further, the operators have to be trained each time which causes a fall in production and an increase in cost. Another example is that the quality is not given the prime focus while buying the raw materials, in particular fabrics. Price is considered the most important factor. As a result, low-quality fabrics are purchased due to which our consumption per unit is increased which results in the overall high cost of the product. It should be kept in mind that the fabric contributes 40–60% of the total cost of an apparel product.

Sales are the third function of the marketing department. Sales teams are responsible for market research, sales forecast, promotions, and selling. It should be kept in mind that the sales teams are only functional in the firms where new products are made.

8.7 Finance/Admin Division

The basic responsibility of a *Finance department* is to manage the finances. In the apparel industry, the payment of an order is made after the goods are received by the buyer. Therefore, the manufacturer has to arrange the finances for the complete order.

Human resource is considered the most important resource nowadays. Companies having better human resource may create a competitive advantage over other companies. The basic responsibility of the *Admin/Human resource* department is to manage the human resource. The responsibilities of the Human resource department are as follows:

- to find out and hire the appropriate human resource for the company
- to train the human resource according to the needs of the company
- to keep the human resource motivated
- to retain human resource

8.8 Manufacturing Division

The manufacturing department is the heart of an organization. The manufacturing department of an apparel company is subdivided into three parts, that is, cutting, sewing, and finishing.

The basic objective of a cutting department is to provide sufficient and quality cutting to the sewing department. The cutting department receives the fabric from the store. Further, they receive a technical package from the marketing. By using the information given in the technical package, they plan how to spread and cut for a particular order. Once they have planned order, they spread and cut. Finally, they number all the bundles and send them to the sewing department.

The basic objective of the sewing department is to stitch. There are multiple operations involved in the manufacturing of an article. The stitching department receives cutting from the cutting department. In addition to that, they receive the technical package from the marketing department. Then they prepare the operation breakdown. After that, they arrange the appropriate sewing machines according to the requirements. After stitching, the stitched garments are sent to the finishing department.

The finishing department is a very important department because this department is responsible for the finished look of the garment. The basic functions of the finishing department are threading, repairing, stain removing, pressing, and packing. The finishing department receives the stitched garments and extra threads (hanging) are cut. Some of the garments get stained during stitching. This problem is mostly faced during the production of white and light-colored garments. Stains are removed from such garments. Repairing is the next step which is conducted on those garments which have got some stitching faults. Further, denim garments may have open seams after washing which are also repaired in the finishing department. After that, garments are pressed and packed according to the buyer's instructions.

More details about the functioning of this department are given in the next section of this chapter.

8.9 Operations Division

Operations division is considered as the brain of the company. The functions of this division are as follows:

- Material planning and sourcing
- Production planning
- Sampling
- Method study
- Time study
- Process improvement
- Quality control and assurance

This division is considered as the supporting department. This division is normally subdivided into several departments, that is, production planning and control (PPC), product development (PD), industrial engineering (IE), and quality.

PPC department is responsible for material planning, material procurement, and production planning. PD is responsible for sampling and new PD.

The IE department is one of the very important assets of a company. The key functions of the IE department are as follows:

- to suggest process improvements
- to simplify the production process
  - to suggest the steps to increase the efficiency
  - to suggest the steps to reduce the overall cost of the process
  - to conduct time study and to calculate the standard minute value and standard allowed minutes for a product
  - to decide the piece rate for the operators

The ultimate goal of the IE department is to simplify the process and reduce cost. Nowadays people are more conscious about the quality of the product they buy. Therefore, an excellent quality department is a necessity to be a good company. The quality department consists of two parts:

- Quality assurance
- Quality control

It is very important to understand the basic concept of quality assurance. In this industry, most of the people have a misconception about quality assurance. Quality assurance is, in fact, everything that one performs to make things right at the very first time. For example, training and machine maintenance are examples of quality assurance. Each employee performs to make sure that the product made is according to the requirements. As a matter of fact, everyone wants to avoid the rework and rejection of products. Both rework and rejection cause different types of losses: material loss and/or time loss. We can also say that the steps for quality assurance are taken before production starts. In this industry, once an order is ready to be shipped, the cartons are opened and checked. This activity is considered as quality assurance which is a misconception. In fact, it is a part of quality control.

Quality control is different from quality assurance. The objective of quality control is to check if the products being produced are according to the given specifications. Their entire job is during the production process. The quality control department has quality inspectors/auditors who check the product specifications at different stages of production. In most of industries, the quality control teams are divided into two parts. First team is responsible for inline quality inspection. Inline quality inspection means that the quality inspectors check the products at different machines randomly. The second team is responsible for the final inspection. Final inspection means the inspection of goods at the end of the production line. The amount of inspection varies from industry to industry. We can reduce the inspection cost by reducing the number of samples to be checked. But it increases the risk of bad products being passed.

The costs incurred on quality control and quality assurance are a part of the cost of quality (CoQ). Most of these costs are the costs of nonvalue-added activities. It is necessary to spend on quality, but the CoQ must stay at a minimum level. Further, we can reasonably reduce the CoQ control by increasing a small amount of CoQ assurance. As a result, the total CoQ may be reduced.

8.10 Clothing Production

Clothing production involves multiple steps which are shown in Fig. 8.4. These steps are further explained as follows:

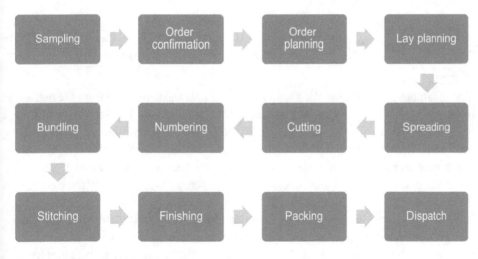

Fig. 8.4: Flowchart of clothing productions.

8.11 Sampling

Sampling is the first step in clothing production. Sampling is normally the responsibility of the sampling department. In some industries, the same department is known as the PD department or research and development department. PD means developing a new product. Unfortunately, most of garment-making organizations in Pakistan do not produce new products. Instead, they reproduce the samples given by the customers. Several types of samples are produced and sent to the customers at different times.

The first sample for a product is known as *Prototype*. The customers want to judge the capability of a manufacturer from these samples. These samples are very important because customers make up their minds to work with the manufacturer after inspecting these samples. Prototype samples are normally made by the sampling department. The approval process of a prototype sample is given in Fig. 8.5.

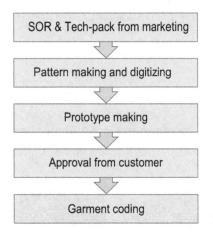

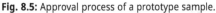

Fig. 8.5: Approval process of a prototype sample.

First, a sample-order request along with a technical package is received from the marketing division. The sampling department makes the patterns and digitizes them (if they have a CAD system). After that the sample is stitched using the available fabrics and sewing thread. Then, the sample is sent to the customer through the marketing division. Customer may accept the sample as it is or suggest some changes. If the customer suggests the changes, the same process is repeated. Once the sample is approved by the customer, the garment is given a unique code so that it may be used in future as a reference.

Salesman sample is the next type of sample after the approval of the prototype sample. The customers display these samples in their stores so that the consumers may know about the upcoming products. In this sample, all the raw materials and accessories used are exactly according to the requirement of the customer. The first step is fabric development and approval. The details of this process are given in Fig. 8.6.

First, fabric swatches are received from the marketing division. The sampling department analyses the fabric, and the fabric characteristics are noted. After that,

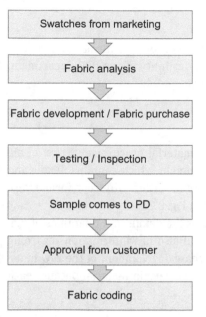

Fig. 8.6: Fabric development and approval process.

either the fabric is developed in-house or it is purchased from outside the firm. After that, the fabric is tested and inspected to make sure that the fabric meets the customer's requirements. Now the developed/purchased fabric sample is sent to the customer through marketing. Customer may accept the sample as it is or suggest some changes. If the customer suggests the changes, the process is repeated. Once the sample is approved by the customer, the fabric is given a unique code so that it may be used in future as a reference. Accessories and sewing thread are purchased during the fabric approval process. Now the salesman samples are prepared and sent to the customer.

Next, the samples are prepared in all sizes provided by the customer. These types of samples are known as a *size set*. The objective of these samples is to check the measurements and shape of the garment before production starts. These samples may be accepted in its form or with minor changes. These samples are made in the sampling room.

Another type of samples is the *pre-production samples*. These samples are normally produced in the production line before the start of production. The customer checks the samples and sends his/her comments. Any changes suggested are made before the start of the bulk production.

Finally, samples are taken during production and sent to the customer. These types of samples are known as *during production samples*. The customer compares these samples with the samples already sent to him.

8.12 Order Planning

Orders are confirmed as a result of sampling. Once an order is confirmed, order planning is started. Planning is considered as the responsibility of the PPC department. There are two basic functions of the PPC department:
- Material planning
- Production planning

In some firms, there is a separate department for material procurement. First of all, they prepare bill of materials (BoM). BoM is a document which contains details and quantity of the materials required to manufacture an order. Different formats are used in different firms. A sample BOM is depicted in Fig. 8.7.

It should be kept in mind that the BoM contains the details and quantities of materials required to complete an order. Sometimes, it also has information regarding the cost of the materials. The next step is the purchase/development of the raw materials. It is better to work with multiple suppliers for a certain material because each supplier has limitations.

Bill of Material					
Purchase Order #: _____ **Customer ID:** _____			**Order quantity:** _____ **Style #:** _____		
Sr.#	Ref.#	Description	Quantity / unit	Total	Grand Total (5% access)
1	Fab-001	Body fabric (meters), 1x1 plain, Blue (180 gsm)	1.5	1500	1575
2	Fab-002	Pocketing (meters), 1x1 plain, white (90gsm)	0.1	100	105
3	Acc-120	Buttons (#)	5	5000	5250
4	Acc-121	Sewing thread (meters)	150	150000 (50 cones)	53 cones
Prepared by: _____ **Verified By:** _____ **Approved by:** _____					

Fig. 8.7: A sample bill of materials.

Vendor assessment is a part of PPC's job. They must see which vendors are better than the others. PPC department keeps in mind three things when buying from a vendor:

- Price
- Quality
- In-time delivery

As a rule of thumb, price is considered the most important, but this is not a good approach. The quality of the material should be given more importance. Only a good quality raw material can make good quality products which can then be sold at a good price. PPC department should judge the importance of these three factors according to the situation.

The second part of order planning is production planning. In some places, it is also called scheduling. PPC department must know the capacity of each of their department. PPC department assigns the work keeping in view the capacity of a department and availability of raw material for an order. Unfortunately, our suppliers are not able to provide the raw material within the time committed, most of the time. That is the reason why the PPC department has to make frequent changes in its plans/schedules. Sometimes firms receive orders above their capacity. In such situations, some of the orders are outsourced.

The contract is known as CMT. According to the contract, we hire a small manufacturer who has the capability of cutting, stitching, and finishing. We place the order with the manufacturer and pay him for his services. In such contracts, quality control is the responsibility of the parent firm.

8.13 Lay Planning

Once the fabric for an order is received, it is inspected by the quality department. If the fabric is rejected due to poor quality, it is sent back to the supplier. In case the fabric is accepted by the quality department, it is issued to the cutting department. The cutting department starts cut-order planning and lay planning. The cutting department first prepares the shade sequence. They arrange all the fabric rolls in a sequence (from lightest to darkest shade). Then lay planning is done keeping in view the total order quantity, sizes, and size ratio.

First of all, the base patterns are received from the sampling department. These patterns are **digitized** with the help of digitizing table. The next step is **grading**. Grading is the process of making patterns of all sizes from the base size. For grading, a rule table is made. Let us try to understand it with a simple example. Figure 8.8 depicts a pattern piece and Tab. 8.1 describes its measurements in four different sizes.

For the pattern piece given in Fig. 8.8, the rule table is given in Tab. 8.2.

Tab. 8.1: Size chart.

Size	Length (cm)	Width (cm)
S	4	2
M	6	3
L	8	4
XL	10	5

P2   P3

P1   P4   **Fig. 8.8:** Pattern piece.

Tab. 8.2: Rule table.

	P1		P2		P3		P4	
	X	Y	X	Y	X	Y	X	Y
S–M	0	0	0	1	2	1	2	0
M–L	0	0	0	1	2	1	2	0
L–XL	0	0	0	1	2	1	2	0

We can make patterns of all sizes by applying the rule table given in Tab. 8.2. Now that we have patterns in all sizes, model making is the next step. A model contains information about the garment parts and their quantity. The model of basic trousers is given in Tab. 8.3.

Tab. 8.3: Model of basic trousers.

S. no.	Garment part	Quantity
1	Front panel	2
2	Back panel	2
3	Yoke	2
4	Back pocket	2
5	Facing	2
6	Watch pocket	1
7	Waistband	1
8	Right fly	1
9	Left fly	1

For one type of fabric in a garment, one model is made. For example, in trousers pocketing fabric is not added in the model as the pocketing fabric is normally dif- ferent from the base fabric. Order making is the next step. An order guides us about how many pieces of each size must be placed on the marker. Table 8.4 describes an order.

Tab. 8.4: An order.

Size	S	M	L	XL
Quantity	1	2	2	1

Here we can see that we will have to place one piece of size S, two pieces of size M, two pieces of size L, and one piece of size XL. The next step is to make a marker breakup. Marker breakup is the plan which contains the information about the total number of markers and the total number of garments of each size on every marker to complete the cutting of an order. A sample of marker breakup for a denim trouser is illustrated in Fig. 8.9. It is worth mentioning here that a half garment marker is usually used in denim trouser manufacturing.

Marker Breakup					
Order #	ABC123				
Customer	XYZ				
Fabric reference	Fab-001				
Total quantity	600				
Size name	S	M	L	XL	Total
Size ratio	1	2	2	1	6
Quantity required	100	200	200	100	
Marker 1(70 plies)	35	70	70	35	
Marker 1 Repeat 1(70 plies)	35	70	70	35	
Marker 1 Repeat 2(60 plies)	30	60	60	30	
Total quantity	1100	200	200	100	

Fig. 8.9: Marker breakup.

Marker making is the next step. The process of placing patterns on the fabric in a way that the fabric utilization is maximum is known as marker making. Marker making is very important because the raw material cost is 50–60% of the total cost in the apparel industry. It means that if we can save 2% of the fabric, we can increase 1% profit margin.

Markers may be of different types which are as follows

- Half garment marker
- Whole garment marker
- Single-size marker
- Multisize marker
- Sectional marker
- Interlocking marker
- Mixed multisize lay

Details of marker types are out of the scope of this book.

8.14 Spreading

Once the marker is ready, information about the marker length is sent to the spreading team. Spreading is the process of superimposing fabric layers to get the required number of garment pieces. One layer of the fabric is known as ply or layer whereas the multiple numbers of layers ready to be cut are known as lay or spread. There are two types of spreading:

- Manual spreading
- Automatic spreading

Fig. 8.10: Spreading/cutting table.

Manual spreading is widely used. Particularly, in home textiles, there is no concept of using automatic spreading. Very few industries use automatic spreading. In manual spreading, it is a two-man team: one on each side of the spreading table. The fabric roll is placed along the width of the table. A spreading table is shown in Fig. 8.10.

There are two modes of fabric spreading, that is, face-to-face and face one way. Figure 8.11 explains the fabric spreading modes.

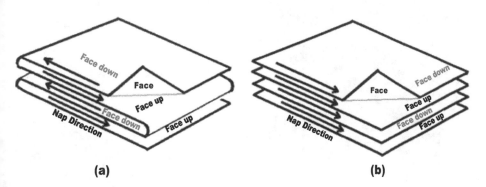

Fig. 8.11: (a) Face-to–face and (b) face one way.

Face-to-face spreading is easier and less time-consuming. It may be used for products which have mirror pieces. In the denim industry, face-to-face spreading is used to make denim trousers. On the other hand, face one way is more time-consuming. But this is the only technique which can be used for directional fabrics. Directional fabric means the fabric having any directional effect or print on its surface. During spreading, shade sequence is kept in mind to avoid shade variations at later stages.

8.15 Cutting

When a lay is ready, cutting is the next process. There are two types of cutting:
–   Manual cutting
–   Automatic cutting

Manual cutting is a widely used method in the industry. A reciprocating knife hand cutter is used for this purpose. Figure 8.12 depicts a fabric hand cutter (Eastman company). The cutter has a base plate which is inserted beneath the lay for cutting. There is a straight knife which reciprocates vertically to cut the fabric. The knife is driven by a motor. Furthermore, there is a handle as well. The operator holds the cutter with this handle and pushes the cutter to cut through the fabric lay.

The other type is automatic cutting in which a computer-operated cutting machine is used. The process is also known as computer-aided manufacturing. An automatic

Fig. 8.12: Reciprocating knife hand cutter.

cutting machine consists of a cutting table and a moving head. Under the cutting table, there is a suction mechanism. After spreading, the lay is dragged onto the cutting table and a polythene bag is placed on the upper side of the lay. Now the suction fans are switched on. The air entrapped between the fabric layers is sucked. Now the lay is squeezed and adhered to the cutting table. The process is necessary so that the fabric cannot move from its place during the action of very high-speed cutting. After that the cutting head is given a command to cut and it starts cutting. An automatic cutting machine (Gerber company) is shown in Fig. 8.13.

Automatic cutting is faster and more accurate than manual cutting. But a huge investment is required to buy an automatic cutting machine. Particularly, in home textiles, there is no concept of automatic cutting. A popular view is that home textile articles are very simple and straight; therefore, automatic cutting is not required.

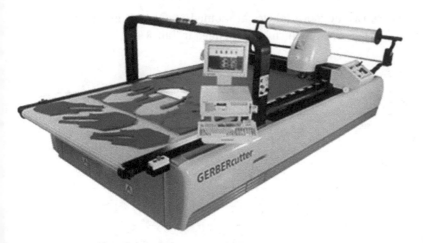

Fig. 8.13: Gerber automatic cutter.

8.16 Numbering and Bundling

Numbering and bundling are the next processes after cutting. Each bundle is labeled. The label contains the information, that is, pattern, size, and number of pieces. The objective of this labeling is to avoid mixing of different sizes during stitching. Now each piece in a bundle is numbered. For example, the topmost ply is given the number 1 and the lowest ply is given the number 30 in a 30-piece bundle. The objective of this numbering is to avoid shade variation during stitching. The basic idea is to attach all the pieces in a garment from the same ply. Let us try to understand it with an example. Imagine we have spread a lay with two fabric rolls having slight shade variation. Now one sleeve comes from one fabric roll and the other comes from the second roll in the same garment. The shade of both sleeves will be different which is considered a fault. Such garment will be rejected by the quality department.

Now the cutting is transported to the sewing department and bundles of all garment parts of a size are placed on a trolley. The bundles are placed on trolleys in a sequence that each trolley contains one bundle of one size. Therefore, we will have the same number of trolleys as we have the number of bundles in a cut. This can be seen from the marker breakup (Fig. 8.9).

8.17 Stitching

The process of joining fabric to make a garment is known as stitching. There are two things involved in the stitching process, that is, stitches and seams.

The line where two or more fabrics are joined together is known as a **seam**. There are six basic seam classes which are explained in Tab. 8.5.

Tab. 8.5: Seam classes.

S. no.	Seam class	Designation	Figure
1	Superimposed seam	SS	
2	Lapped seam	LS	
3	Bound seam	BS	
4	Flat seam	FS	
5	Edge finishing	EF	
6	Ornamental stitching	OS	

Each seam is further divided into subclasses. Superimposed seam is the simplest and most commonly used seam type. Lapped seam is used to make double-sided garments. Bound seam is used in collars and cuffs. Flat seam is rarely used. Bottom hemming is an example of edge finishing. Embroidery is an example of ornamental stitching. A **stitch** is the specific configuration of a sewing thread. There are three methods to make a stitch.

8.17.1 Interlacing

When a thread or a loop of thread passes over or around another thread or loop of another thread, it is called interlacing.

8.17.2 Intralooping

Passing a loop of thread from another loop of the same thread is called intralooping.

8.17.3 Interlooping

Passing a loop of thread from a loop of another thread is called interlooping.

There are six stitch classes which are described in Tab. 8.6. All the stitch classes are based on interlacing, intralooping, and interlooping.

Tab. 8.6: Stitch classes.

S. no.	Stitch class	Description	Examples
1	Class 100	Single thread chain stitch	Blind stitch machine
2	Class 200	Hand stitch	All hand stitches
3	Class 300	Lockstitch	Single-needle lockstitch, double needle Lockstitch, bartack
4	Class 400	Multithread chain stitch	Feedo, two-thread chain stitch
5	Class 500	Overedging	Overlock, safety, and mock safety
6	Class 600	Cover stitch	Flat lock, interlock

All the sewing machines are designed keeping in view the stitch classes. Each stitch class has subclasses as well. Stitching is done by combining different types of seams and stitches.

Trolleys with fabric bundles are received in the stitching department. Each product goes through different sewing operations during stitching. It is necessary to perform these sewing operations in a sequence. There are sewing operations which cannot be performed after the other operations. Therefore, the sewing department makes a list of operations and their sequence first. It is sometimes called operation breakdown/operation sequence. Then the sewing machines are arranged according to the operation sequence. Each machine is run by an operator. One operator performs only one operation on all the pieces of a bundle and then he moves the bundle to the next operation. One by one operations are performed, and finally, all the parts are

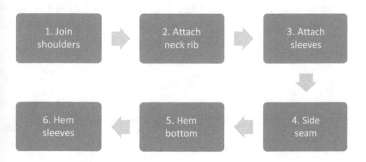

Fig. 8.14: Operation sequence of a T-shirt.

stitched together to make the final product. The operation sequence for a T-shirt is given in Fig. 8.14.

8.18 Finishing

Sewn products are sent to the finishing department. The finishing department takes care of the following functions:
- Trimming
- Stain removing
- Repairing
- Pressing

After stitching, there will be some hanging sewing threads on the finished product. Trimming is the operation of removing these extra hanging threads. Sometimes, finished products get stained during the production process. The finishing department is responsible to remove those stains by using different wetting agents. Some of the sewn products may also have some open seams or other stitching faults. The finishing department repairs such products before packing. The last objective of the finishing department is pressing. The sewn products are pressed to remove wrinkles and enhance the look of the garment.

8.19 Packing

Packing is the last step in manufacturing. This section receives the finished products from the finishing department. Here, the products are packed according to the customer's requirements. Technical package is consulted to get the information about the packing size of the product, carton size, number of pieces in a carton, and their distribution. The size of the carton is a very important factor. The carton is designed in a way that it takes minimum space in the container. After packing, the products are ready to be shipped.

References

[1]   R. E. Glock, and G. I. Kunz, *Apparel Manufacturing: Sewn Product Analysis*, Pearson Prentice Hall, New Dehli, India, 2000.
[2]   World Trade Organization, "International trade statistics," 2012.

Madeha Jabbar, Yasir Nawab

9 Recent Developments in Technical Textiles

Abstract: The textile materials find applications in three broad areas, namely, clothing, home textiles, and technical textiles. Technical textiles deal with the use of textiles (fiber, yarn, or fabric) for technical properties rather than aesthetics. Some typical application areas of technical textiles include buildings and construction, automobiles and transport, agriculture, sports equipment, medical, and industries. The products used in these areas include conveyor belts, tires, seat belts, airbags, filters, and sealing of electronic cables. China and Europe fulfill more than 50% of the global technical textile demand. The global technical textiles market is expected to increase to US $220 billion by 2025.

Keywords: plant fibers, animal fibers, man-made fibers, synthetic fibers

9.1 Introduction

Technical textiles are one of the major categories of various textile products which are primarily defined as the materials and products produced originally for technical properties. The aspect of aesthetics and decorative purposes is ignored in this category of textiles. Technical textiles are utilized in several applications, for example, buildings, construction, vehicles and transport, agricultural uses, making sports equipment, apparel, and medical applications. Industries get the benefit of these products in the form of conveyor belts, tires, seat belts, airbags, filters, sealing of electronic cables, household to aerospace applications. The textile market is a big sector and has been through a long journey starting from cotton, achieving the milestone of carbon in textiles to make super products, and progressing toward nanomaterials development and applications [1].

The technical textile products are not limited to the use of high-performance, high-strength synthetic fibers. It is also about determining the scope of various materials and their potential use in various applications. Natural fibers are a strong candidate for various technical textile products. Their ability to biodegrade is an additional benefit to the ecosystem.

On the other hand, the strength and weight of synthetic fibers provide a competitive edge and a versatile product range for technical end uses. The possibility of end-use applications is practically limitless in the case of synthetic products.

https://doi.org/10.1515/9783110799415-009

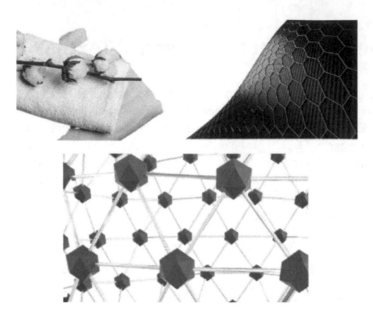

Fig. 9.1: Progress from natural fiber to synthetic and nanomaterials.

9.2 Market and Growth of Technical Textile Products

The global market of technical textiles is approximated to be between US $190 and US $200 billion [2]. The demand for products relevant to technical textiles will greatly increase across Asia-Pacific countries. The main categories include Mobiltech, Indutech, Medtech, and Hometech which comprise more than 50% vol of the total global market. China and Europe fulfill more than 50% of the global technical textile demand of products. India holds 5% of the production of technical textiles. The global technical textiles market is expected to increase to US $220 billion by 2025 [3]. Figure 9.2 shows the category-wise breakdown of the global technical textiles market.

9.3 Circularity in Technical Textiles

In a technical textile circular economy, inputs for textiles are harmless, recyclable, or renewable. Production of textiles does not pose a threat to the health of workers or customers during production or end-use applications. Textiles are able to be utilized for longer time periods. Extending the use phase of textiles maximizes the use of textile products, particularly clothing articles intended to use as for technical applications. Produced textiles should not be left unsold in retail stores or storage areas. And they are also utilized to the fullest of their lives. They are not kept unused in closets or discarded

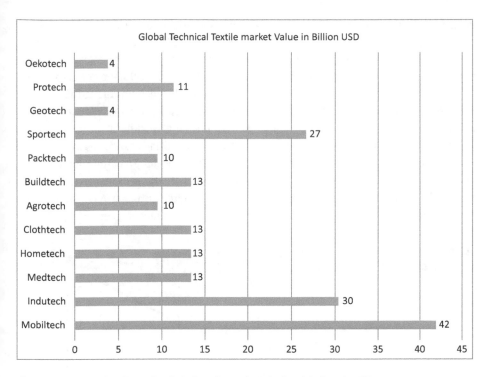

Fig. 9.2: Category-wise share of technical textile products in the global market [3].

while in good condition. Utilization of all manufactured commodities has become the norm both from a production and consumption standpoint. Technical textiles can be recycled and are eventually recycled though it is quite a challenging task. In order to recycle technical textiles that have reached the end of their useful lives are collected and upcycled, as opposed to being sent to landfills or incinerators. Upcycling is prioritized over downcycling in order to preserve the highest potential product value. It is also important to develop (mechanical and chemical) recycling procedures with recycling in mind. The development of a circular economy is the responsibility of businesses, governments, and civil society organizations.

Utilizing high-quality fibers, making them easy to mend, and designing "timeless" styles are strategies for producing durable technical textiles. Utilizing nonhazardous, easy-to-disassemble materials and concentrating on homogeneous fibers rather than complex mixes can facilitate recycling. It will be impractical in the near future for the textile sector to use only recovered materials. Thus, it is important to encourage the market and customers to use fewer textiles and use them for longer durations as well.

Reconsidering consumption means buying fewer products, purchasing used items, supporting sustainable fashion, and extending the life of textiles. Providing environmental and economical benefits wherever used technical textiles are traded can result in major developments in circularity. Collecting and sorting used technical textiles is a labor-intensive operation. All must be thoroughly planned in order to arrive at the correct area and give the correct services. Businesses can only adopt recycled fibers on a wide scale if they are competitive in the market, which supports the expansion of recycled material supply chains. A circular economy for textiles will have complex effects on decent work, shifting employment from agricultural and manufacturing to value chain stages such as repair, resale, and sorting. It has the potential to improve the quality of jobs, especially for informal workers, by improving working conditions, safety, wages, and social security. However, without targeted initiatives from governments, corporations, and civil society organizations, this will not occur. To comprehend the potential socioeconomic effects of increased textile circularity, there is a scarcity of quantitative data; thus, additional research is required to address this crucial information gap. A simple process of circularity of a technical textile product is given in Fig. 9.3.

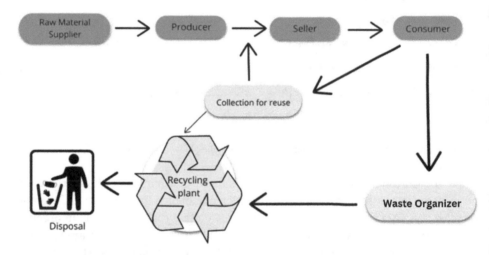

Fig. 9.3: Process of circularity of a technical textile product.

9.4 Conventional Categories of Technical Textiles

Technical textiles are divided into major 12 categories. The use of technical textiles is not limited to only fabric form, but other structures like fiber, yarn, nonwovens, and sheeting for unique applications are also being used in industry. These categories are developed according to the end-use application rather than the form of material used in the products:

1. Medtech – Medical and hygiene products
2. Agrotech – Agricultural, aquacultural, horticultural, and forestry products
3. Buildtech – Construction applications and building materials
4. Mobiltech – Vehicles, automobiles, ships, railways, and aerospace applications

5. Protech – Textiles used for the protection of equipment, persons, and buildings
6. Indutech – Chemical, electrical, and mechanical applications of textiles
7. Hometech – Household textile including furniture, carpets, cushions, and coverings
8. Clothtech – Components of clothing, footwear, and headgear
9. Sport-tech – Sports equipment and clothing
10. Packtech – Packaging products
11. Oekotech – Environment protective products
12. Geotech – Products used in civil engineering and geotextiles

Technical textiles are originated from one or another form of fiber and fibrous materials. The end-use applications include five major structures listed below and shown in Fig. 9.3:
- Fiber
- Yarn
- Woven fabric
- Knitted fabric
- Nonwoven fabrics

Fig. 9.4: Fiber, yarn, woven, knitted, and nonwoven fabrics.

9.5 Conventional Fibers

Often technical textiles are used in the form of yarn or fabrics. The conventional fibers include natural and synthetic materials which are used in technical textiles.

9.5.1 Natural Fibers

Natural fibers have a great place in technical textiles which cannot be ignored. Cotton and silk have been used for technical applications for a long time in human history. Latest technologies are also now being used to modify natural fibers and improve their properties for technical applications. Such processes include fire-retardant finish, soil release, and waterproof finishes. Another important fiber is wool in technical applications. The inherent properties of the ability to trap hair due to convulsions on the surface of fiber have made it an ideal candidate for insulating applications. Recyclability of natural fibers is also an important aspect to integrate natural fibers to make products more environmentally friendly as shown in Fig. 9.4.

Fig. 9.5: Processing of waste natural fibers into compressed wood sheets and bags.

Viscose rayon fiber is a regenerated natural fiber, known for its good water absorbency and is used in various hygiene products. High-tenacity viscose rayon is also utilized as a reinforcement in tires, conveyor belts, and hoses. It also has applications in the aerospace industry. Agricultural and textile industries use it to produce various products such as braided cords, tape, embroidery threads, blanket sheets, and curtains [4]. Viscose rayon filament-based biocomposites are produced and then used as the matrix. The resultant products can be used for applications in medical, automotive, and construction [5]. **Cornhusk and sawdust** briquettes are reported to be used for heating applications to obtain energy from waste materials [6]. **Hemp** fiber is one of the best candidates, which is used in applications where bulk is required. It is also used for heat insulation, sound barrier, and antistatic properties. It is also the second-most cultivated fiber after cotton in the world. Products manufactured from jute fiber include shoe lining, cables, ropes, filter fabrics, packaging materials, and bags. It is used to wrap other materials such as steel tubes and rods. It is one of the most-used fibers in the technical textile sector, that is, Agrotech and Geotextiles. It also has applications to control soil erosion and seed protection. Jute is also used after hybridization with other natural fibers.

A hybrid of **jute** and Luffa's biocomposites and hemp fiber composites have great potential to be used in acoustic applications [7, 8]. The hollow structure of the **hemp** fiber helps to muffle the sound as shown in Fig. 9.6. Hemp is also reported to be used in textile applications to make cords, carpets, canvases, geotextiles, bags, fabrics, apparel,

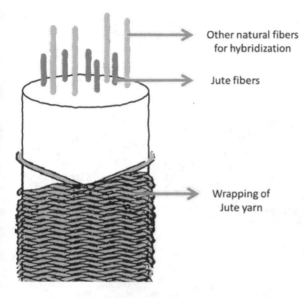

Fig. 9.6: Jute yarn wrapping and fiber hybridization.

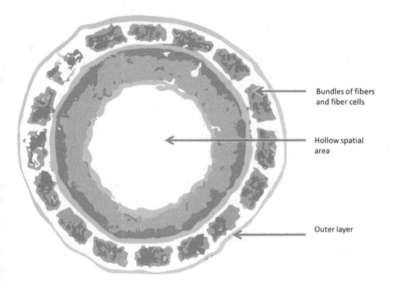

Fig. 9.7: Structure of a stem's interior obtained from hemp.

and apparel accessories. Nonwoven hemp fabrics are used to make composite materials for technical applications.

Hemp, flex, and jute are also under investigation for their low-velocity impact resistance properties. Jute laminates were reported to be a better candidate for impact and energy absorption when compared to hemp laminates. Compression after impact properties of jute-epoxy composites was also better compared to hemp-epoxy composites [9]. Sericulture is an important part of textile silk fiber production. Silkworm cocoons as shown in Fig. 9.7 can be used to obtain silk proteins which are used in natural cosmetic products intended to be used in antiaging, antiwrinkling, and skin-moisturizing applications [10].

Fig. 9.8: Silkworm cocoons ready to be harvested, cornhusk, and sawdust (processed form).

Natural polymers such as collagen, cellulose, silk, keratin, gelatin, and polysaccharides (chitosan and alginate) can be used to make nanofibers utilizing a variety of processes. Collagen is a protein-based material commonly available in animal tissues. It is a needle-thin protein. Proteins are made up of amino acids, which are the basic building components. Amino acids in collagen protein are a precursor to make nanofibers. Small fibers can be stimulated to grow and unite to generate bigger fibers in a biopolymer. This is performed by boosting the material's amino acid content. The increasing level of amino acids causes the tiny fibers to expand and fuse. The final product is robust. Cellulose is a plant-based material that is composed of glucose molecules linked together into chains. Nanofibers are tiny threads or tubes made from cellulose strands in such cases. They have unique physical and chemical properties that make them perfect for various applications including clothing, insulation, medical equipment, and transport.

Silk fibers are unique protein-based materials with great mechanical capabilities that may match high-performance fibers in some ways [15]. Some of the benefits and applications of silk-based materials include flexible electronics, temperature and thermal devices, mechanical power gadgets, detectors, and solar photovoltaic cells. Silk brings several benefits to hybrid power systems, including processability, flexibility, and adaptability. It also provides biocompatibility, high tensile strength, and renewability, making it one of the most all-encompassing materials for use in hybrid devices [16].

9.5.2 Synthetic Fibers

Production of synthetic fibers started a revolution in product design and development for textile applications. It became a breakthrough to obtain high strength, high performance, and intelligent products for technical applications, leading to an era of technical textiles. Fibers which are extensively used in technical textile applications are polyester, polyethylene, polypropylene (PP), nylon, glass, carbon, and metal fibers.

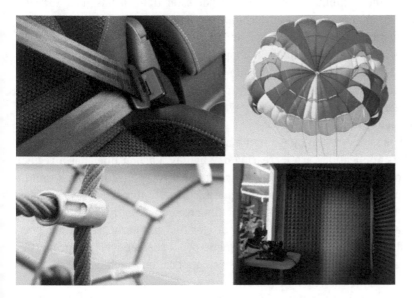

Fig. 9.9: Various end-use applications of synthetic fibers: seat belts, parachutes, ropes, and sound absorption panels.

Polyethylene is a stiff and strong fiber. It is dimensionally stable and acts as a gas barrier. It shows good chemical resistance against greases and acids. It is originally transparent, but coarser fibers appear opaque. It possesses good ultraviolet light resistance and self-extinguishes and provides excellent resistance against the electrical flow and its moisture regain is almost zero. Pertaining to these properties, polyethylene fiber is used in vast applications such as medical textiles, ropes, cables, sailcloth, pressure vessels, boat hulls, impact shields, sports, and fish nets.

Polyester is one of the fibers, which is readily accessible in abundance and cheaper than most of the available synthetic fibers in the world. Polyester fiber's high strength, cheaper rate, and flexibility make it an ideal candidate for various applications. It is also blended with various natural fibers and optimized properties are engineered. Blended materials provide high strength from their polyester content and biodegradability, moisture management, and comfort from the natural fiber content. Polyester is wrinkle resistant with good moisture-wicking properties; thus its blend with cotton is extensively used in garments, hosiery, and upholstery. Woven and knitted fabric domains are heavily shadowed by polyester and cotton blends.

Applications of polyester fiber range from apparel to technical end use such as trousers, wind jackets, dress shirts, socks, cushions, carpets, pillows, duet sets, sewing threads, ropes, tire reinforcements, seat belts, elastic tapes, workwear apparel, belts, tarpaulin, filters, and insulators.

Polyamide started its journey when Dupont made Nylon 6,6. Nylon fiber has greater strength and flexibility as compared to polyester fiber. It has good abrasion resistance, and toughness required for technical applications. The products manufactured from polyamide are lighter in weight and soft.

Nylon dries quickly and maintains its shape, resilient, and resistant to ultraviolet rays and heat. Nylon is also resistant to many common chemicals. Nylon fiber has also vast applications, for example, hosiery, dresses, socks, swimsuits, Winchesters, umbrellas, luggage bags, uniforms, life vests, and parachutes. Car tires and belts are also manufactured from nylon.

Carbon fiber is one of the latest man-made fibers which possesses unbeatable qualities, and its applications are increasing day by day. Carbon fiber has a regular arrangement of graphite crystalline structure which is responsible for its unique properties such as high tensile strength, toughness, stiffness, abrasion resistance, corrosion and chemical resistance, lighter weight, dimensional stability, low thermal expansion, and ability to retain its properties at elevated temperatures. It has good electrical conductivity and great electromagnetic properties as well. Its properties make it an ideal candidate for its use in high-tech applications such as marine, aerospace, defense, civil, medical, and automobile.

PP is the waste product of petroleum-refining processes. It is a rapidly growing synthetic fiber when compared to polyester and nylon. It is majorly used to manufacture industrial products. PP is thermoplastic and exhibits good strength. It has good environmental resistance, low density, and good toughness. Its common uses are as a nonwoven in sanitary and hygiene products, pads, diapers, filtration, and oil-absorbing applications. Hot weather protective clothing also uses PP. It actively transports sweat from the skin to the environment through the cloth barrier and helps to keep skin dry. Products such as ropes, socks, twines, and bale wrappers are manufactured from PP. PP fabrics are also used as a lining material in furniture, coverings, tarpaulins, table clothes and luggage bags, sports equipment and sportswear, and face masks. In civil and mechanical applications high modulus PP is used.

Glass fiber holds the place as one of the first fiber to be used as a reinforcement in modern-day composite products. It consists of extremely fine-extruded fibers of glass. Glass fiber is also used in fabric form (other than industrial applications) where it offers exceptional properties of high strength, fire resistance, thermal insulation, and electrical properties. It has good hardness, resistant to chemicals, stable, and inert. It is transparent by nature. They are used in industrial composites, circuit board printing, flame proofing, and water proofing to make bodies of aircraft, boats, vehicles, tubs, tanks, roofs, pipes, casts, and surfboarding panels' outdoor skins.

Aramid fibers are nowadays among the most-used textiles adapted for technical uses. They can be found in a wide range of industries including automotive, composites, tires, sports, ropes, defense, fire and heat protection, transportation, and aeronautics. It is worth noting that aramid fibers are not dyed very often, if at all, and the color variations are often limited. Colored flame-retardant fibers need significant additional processes and expenses. There are two main categories of aramid fibers: para-aramid and meta-aramids. Bullet-proof coats and cut-resistant gloves are among the many uses for para-aramid.

Meta-aramid (which ranges in shade from white to yellow-orange) is commonly found in personal protective equipment including motorsport suits and firefighter's safety gear either alone or in combination with other fibers like para-aramid to increase the suit's strength value.

Ceramic fibers are mostly created for application areas where the processing or application temperature can reach up to 1,000 C in a caustic and oxidizing condition.

Precursor fibers or thin tungsten core wires are used to create ceramic fibers. Boron and silicon carbide vapor are progressively deposited over the high-temperature precursor. In both tension and compression uses, ceramic fibers exhibit great strength and modulus. The stress–strain curves of unidirectional boron composites are linear until failure and have a modulus of 30 million psi in compression. Prepreg tapes made from ceramic fibers are typically unidirectional solely due to their enormous diameter. Prepregs are stacked and cured in an autoclave for impact-resistant applications such as reinforcing ceramic panels and tiles. To achieve specific goals or higher levels, a variety of different fibers can be mixed with high-performance ballistic fibers.

Typical uses for ceramic fibers include fabrics used to create fire barriers incuding fire safety curtains, engine insulation, and machine/workplace isolation. Land transportation, compartment shielding, marine bulkhead fire coating, and aeronautical engines (fuselage fireproofing) are all examples of thermal insulation. Low-cost glass-fiber building reinforcements; medium- to high-cost load-bearing structural components in transportation applications and composite reinforcement. Flexible fire-barrier fabrics are made from organic fiber mixtures such as aircraft seats, fire barriers, and blockers.

9.5.3 Metal Fibers

Metal fibers are made up of thin strands varying in diameter from 1 to 80 μm. Shredded fibers and needle-felt are two of the many variations. Wet-lay processes can be easily carried out from shredded fibers. Metal fibers can be made from a wide range of metal alloys. Steel, nickel, nickel alloys, and temperature-resisting metals are included. The metal fibers can be utilized on their own for a multitude of scenarios, or they can be converted into other products using various textile manufacturing procedures. It is feasible to spin filament yarns from bundles of endless fibers, but it is also feasible to create spun yarns using classic yarn spinning methods. The fibers are utilized to make hybrid textiles with polyester, cotton, or wool, in addition to pure metal yarns. Knitting, weaving, and braiding can also be used to create textile articles using metal threads. Pure metal fiber yarns can also be employed in these technologies, or material blends can be achieved by using mixed yarns or multiple types of yarns. Metal fibers in a cut state are also appropriate for nonwovens manufacture. Figure 9.9 shows the various products produced from synthetic fibers.

Smart textiles are a key component of the Internet of things which is mainly produced from the integration of metal fibers. This massive digital network connects individuals and a variety of appliances to monitor in real time. Although this technology still appears to be futuristic, it is already assisting many individuals. Arthritis patients with disability in the body parts and who are suffering from arrhythmia are able to continue to live at home due to the integration of smart textiles into their lives. Trackers on his moving body parts give data to online software to ensure that the patient is doing his physical therapy properly. If the patient falls, sensors in the carpet inform his/her contacts. A heartbeat can be monitored 24 h a day via a heart rate monitor. To avoid bedsores, mattresses are connected to the software and internet. Smart fabrics have a wider range of applications beyond healthcare as well. Firefighter suits, RFID tags, and fiber optic garments all employ them for heat sensing. Stainless steel filaments and other metal alloys are used to create conductive yarns. Even after several washings, these yarns keep their performance and flex life.

9.6 Technical Fibers

The ratio of a fiber's surface area to its volume has a direct impact on its behavior as well as its functional capabilities. Because increased fiber fineness results in increased specific surface areas, one may generally anticipate greater liquid absorption and faster drying times. This is because increased fitness leads to the increased specific surface area. These fine fibers would also require less dye or impregnator concentration in order to get the desired shade or quantity, which would contribute to a reduction in the overall cost of production.

Fibers with diameters of less than 100 nm are referred to as nanofibers. Drawing, template synthesis, phase separation, self-assembly, and electrospinning are only some of the processing methods that have been utilized to manufacture nanofibers. Other methods include phase separation and self-assembly. Electrospinning has emerged as the method of generating nanofibers that is the least complicated, has the greatest degree of practicality, and can be done at a reasonable cost.

Thus, finer fibers are more flexible [11], have better drape, and thus feel smoother to the touch and are easier to mold. The molecular orientation and increased intermolecular

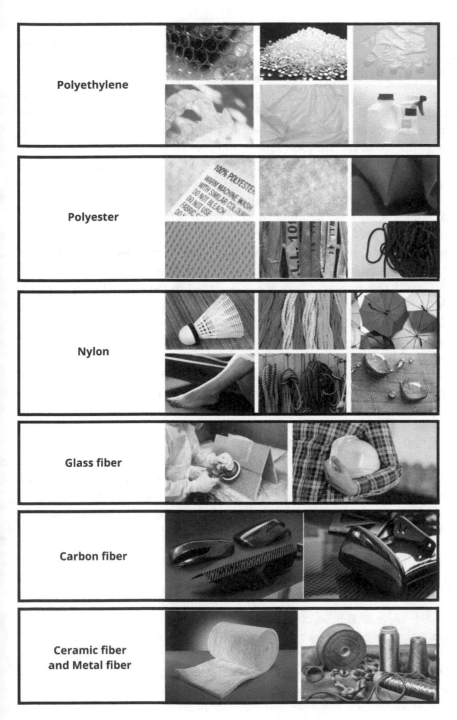

Fig. 9.10: Technical textile products produced from various synthetic fibers.

tensions that exist within these fibers contribute to their increased strength. As a result, these fibers have an inherent advantage over their more coarsely textured counterparts. Due to their outstanding qualities, nanofibers are becoming a viable alternative resource for a variety of materials. They are low cost, have low density, high porosity, and high energy. Nanomaterials are emerging as substitutes for their original materials with respect to the passage of time. Their unique properties enable nanofibers for innovative applications. Fibrous structures are employed as a cost-effective alternative for numerous nonrenewable resources. They have been employed in everyday items like mobile phones, cells, batteries, and filtration barriers, among other things. Coconut fiber is used to make cellulosic microwhiskers [12]. Sea algae are also used to make nanofibers [13]. Chemical and mechanical approaches are used to manufacture nanofibers. Nanofibers made from natural and synthetic polymers are truly nature's best friends.

Technical fibers are divided into two types: conventional and nonconventional. Natural, regenerated, synthetic, high-strength and high-modulus fibers, chemical-resistant, and flame-resistant fibers are all included in these broad categories. Extremely fine and novel fibers, mainly microfibers and fibers with unusual qualities, are referred to as ultrafine and novelty fibers. Civil engineering, automobile, aerospace, medical, hygiene, and defense are all possible applications of nanofibers.

Nanofiber is used in a wide range of applications including automotive, pharmaceuticals, and electronics. Nanofiber is also often used in technical textiles because of its unique properties. These properties make it resistant to wear, flexible, and strong.

9.6.1 Resources of Technical Fibers

Nanofibers are made from various types of materials and they exhibit different physical properties. Nanofibers are made from natural polymers, synthetic polymers, carbon-based materials, metals, ceramics, semiconductor materials, and composite materials. Each of these materials has its own significance and uses [14].

9.6.2 Advanced Fibers

Despite their traditional function as an essential unit cell within textile materials, fibers now play a wider role than ever before. This expanded involvement is due to better understanding of their potential benefits in combination and composite contexts, as well as their declining size association, which improves performance qualities, flexibility, and broader surfaces of interaction. The millimeter was once the most widely accepted smallest unit of measurement for fibers in terms of size for monofilaments, but this generality has shifted to micrometer scales in the last three decades, with the nanometer scale now becoming the dominant measurement and exposition for such materials. Recent developments include auxetic fibers, piezoelectric fibers, photovoltaic fibers, and shape memory fibers. Auxetic materials are best defined as materials which break the rules of physics by expanding when they are stretched. On the other hand, the so-called auxetic effect, which may also be seen in nature as well, is fascinating for a variety of different applications. It can even be utilized to produce innovative applications and even cure injuries and tissue damage. Delivery of drugs, wound repair, tissue regeneration, barrier fabrics, filtration, absorbers, capacitors, detectors, storage, and entrapments are some of the potential applications for this material [15, 16].

The term "piezoelectric effect" refers to the capacity of certain materials to generate an electric charge in response to an externally applied mechanical stress. In textiles, it has led to the development of the revolutionary concept of wearable energy harvesting applications which require piezoelectric textile fibers. While the majority of research has focused on fibers made of polyvinylidene fluoride, fibers from PP are also being researched. Furthermore, the piezoelectric behavior of the fibers has been defined by the voltage generated by the fibers when tested under open circuit conditions mostly [17, 18].

A solar cell is a device that directly converts sunlight into electricity and is also known as a photovoltaic cell. Solar cells are becoming more common. Some photovoltaic cells have the ability to convert artificial light into an electric current. On top of optical fiber, a dye-sensitized solar-cell structure can be deposited to create a specific kind of photovoltaic fibers. These fibers can then be used to generate electricity. Silicon is by far the most common material used in the production of these materials. Optical fibers made of silica or optical fibers made of plastic can both be referred to as "substrate" in some contexts. A fiber such as this one is able to convert the light modes that are travelling through the modified cladding into an electrical signal. This is made possible by the fact that the cladding is modified [19, 20].

A type of material known as a shape memory polymer, or SMP for short, is one that has the capability to recognize and respond to changes that occur in the environment in which it is located. Its behavior is comparable to the intelligent reflection of living things. One of the initial concepts for an industrial application of SMPs was in robotics, where they may provide existing robot systems with a variety of advantages over their capabilities. This was one of the original ideas for an industrial application of SMPs. The use of shape-memory polymers, also known as SMPs, makes it possible to manufacture photonic gratings that are both functional and responsive. This is due to the fact that SMPs have the ability to modify their shape. SMPs are able to respond to environmental stimuli such as an electrical charge or heat by morphing into new forms and adjusting to the conditions of their surroundings, much like intelligent robots are able to do. One of the medical specialities that have profited from the implementation of this technology is orthopedic surgery. Anticounterfeiting labels have been devised that, when exposed to particular chemicals, reveal a graphical symbol or code. These labels can be read only by authorized personnel. There is a chance that labels that serve multiple purposes will make it more difficult to pass off counterfeit products. As soon as the shape memory film is heated, its 3D pattern will be freed so that it can be embossed or vanish in only a few seconds in an irreversible manner. Label substrates or face stock made of shape memory film can be utilized for anticounterfeiting, trademark protection, tamper-evident seals, antipilferage seals, and other applications similar to these uses when using shape memory film [21–24].

9.7 Applications of Technical Textiles

Textiles can be found virtually anywhere; they are worn on the human body for protection and self-expression; they are used as decoration and comfort elements in homes, offices, hospitals, hotels, or public buildings; they are used as interior components in automobiles, buses, trains, ships, and aeroplanes; they are used as structural elements for tents, roofs, and bridges; they are used as reinforcements for roads and dikes; however, textiles can also be found in sports and outdoor activities. In spite of the fact that the textile industry is often seen as a conventional sector, it has emerged in recent years as one of the most important testbeds for creative business strategies. This is despite the fact that the textile industry has been around for centuries. New market standards, which can be accomplished through process innovations and which, on the one hand, cut costs and, on the other hand, allows one to differentiate themselves from other competitors, have emerged as a highly important factor in the environment of intense competition. These new market standards can be achieved through process innovations [25].

Fabrics need to be viewed not only as a surface, to be interpreted graphically, but also as a substance in all intents and purposes, with their intrinsic structure and performance. This is because fabrics have their inherent structure and performance. Considering the state of the world we live in today; it is essential to rethink our attitude to textiles. Within the realm of technical textiles, there are a wide variety of subindustries and products, the majority of which can be classified as highly technological. When it comes to these markets and goods, the end-user has specific expectations, and the cost is no longer the only element that is considered during the design process. When it comes to cutting-edge textiles, the sector is growing at a breakneck pace, and a large number of new product and application innovations are now in the process of being developed. The progression of technology, which interestingly integrates human science, materials science, and information technology, does allow foreseeing positive perspectives in the approach toward the development of new products and applications [26].

These perspectives can be seen in the ability to foresee positive outcomes in the process of creating new products and applications. The technological assets in the field of specialized applications are those that meet the highest standards for both performance and comfort as well as those that guarantee an improved quality of life. There are currently fabrics available for purchase that may reduce the likelihood of certain risks (e.g., antibacterial, mite-proof, insect-proof, odorless, flame-retardant, soil-resistant, anti-UV, and antielectromagnetic radiation). Other fabrics are particularly effective at performing their designated roles (e.g., heat-regulating, with new visual features, providing cosmetic-medical effects). Textiles were almost solely consigned to the duty of interior decorating in the past; nevertheless, in modern times, they are quickly becoming an intrinsic part of the structures in which they are utilized. Because they have improved performance characteristics in terms of their strength-to-weight ratio, durability, flexibility, insulating and absorption properties, and resistance to fire and heat, they are able to replace more conventional building materials such as steel and other metals, wood, and plastics. Because of these qualities, the performance of these materials is superior to that of conventional building supplies.

Some instances of inventive uses of textiles include the following: The skin functions as both a barrier and a protective barrier for the human body, making it the primary line of defense and primary barrier in the human body against the outside world. Additionally, it plays a significant role in the transfer of energy (in the form of heat) and substance (in the form of fluids and gases such as water, oxygen, and others) between the body and the environment. This transfer of energy takes the form of heat. This transfer of substance takes the form of fluids and gases. The protective function of an individual's natural skin has traditionally been augmented using clothes, which serve as a substitute for a second skin and work as an additional layer of protection. Despite this, additional layers of protection frequently have a detrimental impact on the ability of human skin to conduct heat and moisture exchange, sometimes to an extreme degree as is the case with full-body armor, firefighter uniforms, diving suits, and other garments that are very similar to one another. The development of clothing that is both useful and smart or intelligent is a creative response to such limitations as they have been imposed [27].

Products that emphasize a single or several specialized functions are referred to as "functional clothing." Some examples of these functions are high degrees of insulation, resistance to water or fire, breathability, and resistance to wear and tear, amongst others. The idea of "smart clothing," which refers to products that may perform their tasks in a manner that is more adaptive in response to cues from the user or the surrounding environment, takes the concept of (multi) functionality to the next level. For instance, the following are some of the capabilities of smart clothing: When we talk about "protective clothing," we are referring to garments and accessories that are intended to screen individuals from potentially hazardous chemicals, procedures, or occurrences while those individuals are engaged in their professional or recreational activities. In addition to this, it includes clothes whose major function is to provide protection from other individuals in the workplace, the environment, or all of these (as for cleanroom garments). The need for protective gear is influenced not only by increasing rates of industrialization but also by a growing consciousness of the need to comply with health and safety regulations as well as cleanliness regulations. In other words, the need for protective gear is not only influenced by rising rates of industrialization but also by rising rates of industrialization. The rising rate of violent crime, in conjunction with an increase in the number of military operations, has resulted in an increase in public spending that is targeted at lowering the number of injuries sustained by law enforcement, civil defense, and military personnel. This increase in public spending is targeted at lowering the number of injuries sustained by law enforcement, civil defense, and military personnel. Even though there is some debate regarding whether recreational applications truly belong in the category of industry, it is standard practice for many sports and recreational activities to make use of textiles that are designed for industrial settings. This includes things like uniforms, equipment, and apparel. Most of this gear needs to be able to handle heavy and repetitive loads, impacts, stretching, inclement weather, and other tough conditions without cracking or ripping. It also needs to be able to withstand extreme temperatures.

Even though there is some debate regarding whether recreational applications truly belong in the category of industry, it is standard practice for many sports and recreational activities to make use of textiles that are designed for industrial settings. This includes things like uniforms, equipment, and apparel. Most of this gear needs to be able to handle heavy and repetitive loads, impacts, stretching, inclement weather, and other tough conditions without cracking or ripping. It also needs to be able to withstand extreme temperatures. Employees of factories and emergency response services, construction laborers, civil servants, and members of the armed forces, amongst many others, are required to wear personal protective equipment, also known as PPE, to work safely in the hazardous environments that are inherent to their jobs. Industrial textiles are utilized in a wide variety of applications, including those found in the transportation industries of aviation, automotive, and maritime transportation. The requirements of these many industries for support, filtration, and safety can all be met by textiles to a significant degree. Medical textiles are required to meet high regulatory criteria to guarantee that they are safe for use in direct patient contact. This is done to eliminate any potential for infection. Fabrics frequently need to be antimicrobial, resistant to radiation, and moisture-repellent and have a variety of other features to guarantee the wearer's safety. This is because radiation can cause the growth of microbes [28].

References

[1] C. Byrne, Technical textiles market – an overview, Handbook of Technical Textiles, pp 1–23, Jan 2000. doi: 10.1533/9781855738966.1.

[2] India An Emerging Market & Global Manufacturing Hub for Technical Textiles | Wazir. https://wazir.in/india-an-emerging-market-global-manufacturing-hub-for-technical-textiles (accessed Apr. 13, 2022).

[3] Y. Nawab, "Production and Export of Technical Textiles: Harnessing the Potential in Pakistan." https://wwfasia.awsassets.panda.org/downloads/high_res_technical_textile_report_interactive_.pdf (accessed Apr. 13, 2022).

[4] S. L. Agign, and P. Potluri, *Handbook of Technical Textiles, Technical Textile Applications*, 2nd ed. vol. 2. Elsevier, 2016. doi: 10.1016/B978-1-78242-465-9.00001-X.

[5] V. Gorade, B. Chaudhary, O. Parmaj, and R. Kale, "Preparation and Characterization of Chitosan/viscose Rayon Filament Biocomposite," 2020, https://doi.org/10.1080/15440478.2020.1764442, doi: 10.1080/15440478.2020.1764442.

[6] O. A. Akogun, M. A. Waheed, S. O. Ismaila, and O. U. Daeo, "Physical and Combustion Indices of Thermally Treated Co Hnhusk and Sawdust Briquettes for Heating Applications in Nigeria," 2020, https://doi.org/10.1080/15440478.2020.1764445, doi: 10.1080/15440478.2020.1764445.

[7] Y. Saygili, G. Genc, K. Y. Sanliturk, and H. Couk, "Investigation of the Acoustic and Mechanical Properties of Homogenous and Hybrid Jute and Luffa Bio Composites," 2020, https://doi.org/10.1080/15440478.2020.1764446, doi: 10.1080/15440478.2020.1764446.

[8] J. Liao, S. Zhang, and X. Tang, "Sound Absorption of Hemp Fibers (Cannabis Sativa L.) Based Nonwoven Fabrics and Composites: A Review," 2020, https://doi.org/10.1080/15440478.2020.1764453, doi: 10.1080/15440478.2020.1764453.

[9] S. Patil, D. M. Reddy, N. J, S. S. Swamy, V. P, and G. Venkatachalam, "Low-Velocity Impact and Compression after Impact Properties of Hemp and Jute Fiber Reinforced Epoxy Composites," pp. 1–16, Apr 2022, https://doi.org/10.1080/15440478.2022.2057383, doi: 10.1080/15440478.2022.2057383.

[10] J. Grześkowiak, and M. Lochynska, "Possibilities and Conditions of Developing Mulberry Silkworm Rearing in Poland, Especially for Purposes Related to the Production of Natural Cosmetics Based on Silk Proteins Obtained from Silkworm Cocoons," pp. 1–10, Apr 2022, https://doi.org/10.1080/15440478.2022.2063220, doi: 10.1080/15440478.2022.2063220.

[11] S. K. Bhullar, et al., Design and fabrication of eoxetic PCL nanofiber membranes for biomedical applications, Materials Science and Engineering: C, vol. 81, pp. 334–340, Dec. 2017. doi:10.1016/J.MSEC.2017.08.022.

[12] M. F. Rosa, et al., Cellulose nanowhiskers from coconut husk fibers: Effect of preparation conditions on their thermal and morphological behavior, Carbohydrate Polymers, vol. 81, no. 1, pp. 83–92, May 2010. doi: 10.1016/J.CARBPOL.2010.01.059.

[13] H. Gao, et al., Fabrication of cellulose nanofibers from waste brown algae and their potential application as milk thickeners, Food Hydrocolloids, vol. 79, pp. 473–481, Jun. 2018. doi: 10.1016/J.FOODHYD.2018.01.023.

[14] Kenry, and C. T. Lim Nanofiber technology: current status and emerging developments, Progress in Polymer Science, vol. 70, pp. 1–17, Jul. 2017. doi: 10.1016/J.PROGPOLYMSCI.2017.03.002.

[15] R. Cranberry, B. Holschuh, and J. Abel, Experimental investigation of the mechanisms and performance of active auxetic and shearing textiles, Proceedings ASME Conference Smart Materials, Adaptive Structures and Intelligent Systems, Vol. 2019, Sep 2019. doi: 10.1115/smasis2019-5661.

[16] M.U. Nazir, K. Shaker, R. Hussain, and Y. Nawab, Performance of novel auxetic woven fabrics produced using helical auxetic yarn, Materials Research Express, vol. 6, no. 8, pp. 085703, 2019. doi: 10.1088/2053-1591/ab1a7e.

[17] S. Waqar, L. Wang, and S. John, Piezoelectric energy harvesting from intelligent textiles, Electronic Textiles: Smart Fabrics and Wearable Technology, pp. 173–197, Jan 2015. doi: 10.1016/B978-0-08-100201-8.00010-2.

[18] D. Matsouka, S. Vassiliadis, and D. V. Bayramol, Piezoelectric textile fibres for wearable energy harvesting systems, Materials Research Express, vol. 5, no. 6, p. 065508, Jun 2018. doi: 10.1088/2053-1591/AAC928.

[19] M. Peng, B. Dong, and D. Zoo, Three dimensional photovoltaic fibers for wearable energy harvesting and conversion, Journal of Energy Chemistry, vol. 27, no. 3, pp. 611–621, May 2018. doi: 10.1016/J.JECHEM.2018.01.008.

[20] M. Toivola, M. Ferenets, P. Lund, and A. Harlin, Photovoltaic fiber, Thin Solid Films, vol. 517, no. 8, pp. 2799–2802, Feb 2009. doi: 10.1016/J.TSF.2008.11.057.

[21] Y. Xia, Y. Liu, and F. Zhang, Nano/micro structures of shape memory polymers: From materials to applications, Nanoscale Horizon, vol. 5, p. 1155, 2020.

[22] L. Wang, F. Zhang, Y. Liu, and J. Leon, Shape memory polymer fibers: Materials, structures, and applications, Advanced Fiber Materials, vol. 4, no. 1, pp. 5–23, Apr 2021. doi: 10.1007/S42765-021-00073-Z.

[23] J. Hu and J. Lu, Shape memory fibers, Handbook of Smart Textiles, pp. 1–21, 2014. doi: 10.1007/978-981-4451-68-0_3-1.

[24] R. Tonndorf, D. Aibibu, and C. Cherif, Thermoresponsive shape memory fibers for compression garments, Polymers (Basel), vol. 12, no. 12, pp. 1–13, Dec 2020. doi: 10.3390/POLYM12122989.

[25] A. Rasheed, Classification of technical textiles, Topics in Mining, Metallurgy and Materials Engineering, pp. 49–64, 2020. doi: 10.1007/978-3-030-49224-3_3.

[26] B. J. McCarthy, An overview of the technical textiles sector, Handbook of Technical Textiles: Second Edition, vol. 1, pp. 1–20, Jan 2016. doi: 10.1016/B978-1-78242-458-1.00001-7.

[27] R. G. Revaiah, T. M. Kotresh, and B. Kandasubramanian, Technical textiles for military applications, vol. 111, no. 2, pp. 273–308, Feb. 2019. https://doi.org/10.1080/00405000.2019.1627987, doi: 10.1080/00405000.2019.1627987.

[28] A. Richard, and S. C. Anand, *Handbook of Technical Textiles. Vol. 2, Technical Textile Applications*. Amsterdam: Elsevier, p. 453, 2016.

Sheraz Ahmad, Faheem Ahmad

10 Textile Testing

Abstract: Testing is the method or series of methods carried out to find the quality of a raw material or a final product; in the case of textiles, the product could be yarn, fabrics, and so on. The testing of textiles from fibers to the final product plays a key role in making and achieving the required properties of a textile product. Moreover, the characterization of textiles by various testing techniques enables to optimize the textile manufacturing process. The testing process for textile products is a costly and time-consuming operation; hence, it must add some benefits to the process of preparation, and the quality of the final product must be enhanced. In this chapter, various testing techniques to test fibers, yarns, and fabrics are discussed.

Keywords: yarn testing, fabric testing, mechanical properties, comfort properties

10.1 Introduction

The quality of a textile product is determined before its large-scale production. The testing provides the quality and information about various properties of the textile product. The testing is usually described as the standard method performed to determine the properties and performance of the product. Textile testing is normally categorized into two types: regular process testing and quality assurance testing. Textile testing is also classified based on data and testing techniques employed to test the product. Moreover, textile testing is also divided into destructive and nondestructive types of testing. The nondestructive testing is the method of inspection/evaluation or measurement without damaging the sample. The fabric drape testing, air permeability test, and evaluation on the Kawabata system are the nondestructive testing methods for textiles. The testing specimen is damaged and ruptured in destructive testing to characterize the various properties of the textile product. Common examples of destructive testing are tensile strength testing, elongation testing, and tear testing of textile fabrics. All these types of testing are performed during and after the manufacturing of the finished textile product to assure various quality parameters.

Quality could be termed as customer's satisfaction; a good quality product means that it will fulfill all the purposes for which it has been produced [1].

https://doi.org/10.1515/9783110799415-010

10.2 Quality Control

Quality control is a continuous and regular control of the parameters which affect the quality of the final product. The quality of a textile product is determined by keeping an eye on the final application. The quality of a textile product is judged on a scale of various parameters which can vary from product to product and process to process. It comprises planning, raw data compilation, investigation, and implementation. The objectives of quality control are as follows:
- To select the required textile raw material
- To produce the required quality product
- To fulfill the customer's required standards
- To reduce the production process cost
- To minimize process wastage
- To get maximum profit at the minimum possible cost

10.3 Standardization of Testing

When a textile material or product is tested, its results must fulfill both explicit and implicit requirements. The explicit requirements from the tests are either how the product will perform during its life cycle or how it will meet the required specifications. The core purpose of testing is that it must be reproducible; it means that if the same material is tested under the same conditions in a different laboratory, or at another time, and by another operator, it should yield the same results. However, the testing results of textile materials are not expected to be the same every time. The factors affecting the reproducibility of results are as follows:
- *Variation in the material*: It could be minimized by the proper selection of a representative sample from the material being tested by using some statistical tools.
- *Variations imparted by the test method*: These variations are caused by conditions under which the testing is being held like speed, pressure, gauge length, temperature, and relative humidity. These variations could be minimized by specifying the standard written test methods for testing. For this purpose, organizations like ISO (International Organization for Standardization) are working to build internationally accepted standard test methods.

10.4 Test's Repeatability and Reproducibility

The precision of any testing method is evaluated through redundant testing of the same material between various testing laboratories and within the same laboratory by using identical testing instruments. These testing results are compared, and the reproducibility/repeatability of various laboratories is computed. The variations between testing laboratories are always higher than the variations observed within a single laboratory.

The repeatability is defined in the following section.

10.4.1 Repeatability

(a) Qualitatively: It is the accuracy between consecutive test results collected through identical methods for the same material at the same testing conditions.
(b) Quantitatively: It is described that a value below which the difference among results of two tests collected under the abovementioned conditions might expect to fall with a discrete value of probability [1].

10.5 Sampling

It is not necessary to perform testing of the complete material and it is practically impossible as well, due to time and cost factor. Also, some tests are of destructive type, in which no material remains after testing. Due to this, we use model/symbolic samples from the bulk material for testing. A major objective of sampling is to make an unbiased sample representing the whole.

The terms that are being used in sampling are described further.

10.5.1 Consignment

It can be described as the quantity of a material which is transported together or at the same time. Every consignment may have a single or a number of lots.

10.5.2 Lot for testing

Normally, every container of a material having one specific variety and amount which is delivered to the customer is considered as a test lot. It is presumed that the whole material is uniform, and all parameters of the material are checked. The materials or final products are usually rejected on the basis of variations in the single lot.

10.5.3 Laboratory Sample

Samples of the material which are used to perform testing in the laboratory are considered as laboratory samples. Random sampling procedures are adopted to derive laboratory samples from the lot.

10.5.4 Test Specimen/Sample

The test specimen presents a portion of the material that will actually be utilized for testing, which is taken from a laboratory sample. Several test specimens are tested to get reliable test results. Moreover, variation in these selected test specimens is of great interest to assure the quality of the product.

10.5.5 Package

An elementary unit within each container in the consignment is known as a package. The package might be fibers, sliver cans, bobbins, cones, fabric rolls, and so on [1].

10.6 Moisture Content and Moisture Regain

The quantity of moisture present in textile fibers is indicated in the form of its moisture content or moisture regain. In the textile industry, the most used term is moisture regain.

"Moisture regain" is presented as the weight of water in textile fibers described as the percentage of the oven dry weight:

$$\text{Moisture regain} = (100 \times W)/D \qquad (\text{expressed in \%})$$

"Moisture content" is defined as the weight of water in textile fibers described as the percentage of the total weight:

$$\text{Moisture content} = (100 \times W)/(D + W) \quad (\text{expressed in \%})$$

where D is the dry weight and W is the weight of the absorbed water [1].

10.7 Standard Atmospheric Conditions for Testing

The properties of textile materials are significantly changed as the moisture content changes, especially for natural fibers, so it is the basic requirement to standardize the temperature and relative humidity in which any testing procedure is executed. For this purpose, standard conditions of temperature and relative humidity have been set for the testing environment; that is, a relative humidity of 65% ± 4 at a temperature of 20 °C ± 2. By maintaining these conditions, the repeatability and reproducibility of testing are more likely to be enhanced [2].

10.8 Testing of Fibers

Testing of raw materials is very important, and in the case of spinning, raw materials could be fibers, so it is necessary to find out the complete parameters of fiber that will help in the process of spinning. Therefore, the selection of textile fibers is usually done after performing various fiber quality tests. The properties of the yarn and fabric are directly influenced by the properties of fibers from which yarn and fabric are made. For testing of fibers, mostly used instruments are USTER HVI (high-volume instrument) and USTER AFIS (advanced fiber information system). The testing of textile fibers is done on the basis of certain testing standards by using these instruments. The test results are given in Table-1.

Tab. 10.1: Comparison of parameters measured by USTER HVI and USTER AFIS.

A comprehensive explanation of parameters evaluated by USTER HVI	A comprehensive explanation of parameters evaluated by USTER AFIS
UHML: 1) UHML could be defined as the average length of the longer (upper) half fibers in the sample checked. Its unit will be in inches or millimeters.	1) Nep (count per gram): It is the number of entanglements of fibers in the cotton sample.
2) Meanlength: This is the average length of all fibers in the sample being tested. Its unit will be in inches or millimeters.	2) Nep size: It is the size of the total Neps in the cotton sample and expressed in micrometers.
3) UI: It is also defined as length uniformity, and it is the ratio between the average length and the UHML of the fibers. It is described in percent.	3) Seed coat Neps (count per gram): It is the number of fragments of cotton seeds in the sample that still have some fibers attached.
4) SFI: This is the amount of fibers in percent that is less than 0.5 inch (12.7 mm) in length.	4) Seed coat Nep: It is the size of the total seed coat Neps in the cotton sample and expressed in micrometers.
5) Strength: It is defined as the force in grams required for breaking a bundle of fibers having one tex unit size. It is expressed in gram/tex.	5) Mean length by weight: It is the average fiber length of all the cotton fibers in a cotton sample computed on a weight basis and expressed in inches or millimeters.

Tab. 10.1 (continued)

A comprehensive explanation of parameters evaluated by USTER HVI	A comprehensive explanation of parameters evaluated by USTER AFIS
6) Elongation: The average length of distance to which the fibers extend before breaking. It is expressed in percent.	6) UQL by weight: It is the length of the longer 25% of all fibers in a cotton sample on weight basis and expressed in inches or millimeters.
7) Moisture: Amount in percent of water (H_2O) which is present in the sample being tested.	7) Short fiber content by weight: The percent of all fibers in a cotton sample that is shorter than 0.5 inches or 12.7 mm on weight basis.
8) Rd: This value expresses the whiteness of the light reflected by the cotton fibers. It corresponds to Rd in the Nickerson–Hunter color chart.	8) Mean length by number: It is the average fiber length of all cotton fibers in a cotton sample computed on a number basis and expressed in inches or millimeters.
9) Yellowness (+b): It is a measure of the yellowness of the fiber and is based on the Nickerson–Hunter scale. Cotton ranges from 4 to 18.	9) UQL by number: UQL is defined as the length of the longer 25% of all fibers in a cotton sample on the number basis and expressed in inches or millimeters.
10) Color grade (C grade): Reflectance (Rd) is used in conjunction with the yellowness (+b) to determine the instrument-measured color grade of cotton.	10) Short fiber content by number: The percent of all fibers in a cotton sample that is shorter than 0.5 inches or 12.7 mm on the number basis.
11) Trash count/amount: It is a measure of the trash of cotton. It is the number of trash particles measured on the surface of the sample.	11) 5% length by number: It is the length of the longer 5% of all fibers in a cotton sample and expressed in inches or millimeters.
12) Trash area: The value is also a measure of the contamination of cotton. It indicates the contaminated area in comparison to the total area of the sample measured.	12) Fineness: It is the mean fiber fineness (weight per unit length) in millitex. One thousand meters of fibers having a mass of 1 mg equals 1 mtex.
13) Trash grade: Classer's leaf grade depends upon trash area (%) and ranges from 1 to 8.	13) Maturity ratio: It is the ratio of fibers having 0.5 (or more) circularity ratio divided by the amount of fibers having 0.25 (or less) circularity.
14) Micronaire (Mic.): It is the fiber fineness. A sample of fibers having constant weight is tested by passing air through it and the drop in pressure is recorded then. It is expressed in μg/inch.	14) Immature fiber content: It is defined as the percentage of fibers in a sample having a cell wall thickness covering less than 25% of the full area.

Tab. 10.1 (continued)

A comprehensive explanation of parameters evaluated by USTER HVI	A comprehensive explanation of parameters evaluated by USTER AFIS
15) Maturity index: Is a relative value calculated from other HVI measurements, such as micronaire, strength, and elongation. It indicates the degree of cell wall thickness in a sample.	15) Dust count per gram: It describes the smaller particles from the plant and simply dirt from the cotton field that sticks with the plant (particle size <500 μm).
	16) Trash count per gram: It is the general term for larger impurities containing particles from the cotton plant itself and other plants contaminating the cotton fields (particle size >500 μm).
	17) Total trash count per gram: All particles removed by AFIS fiber individualizer, regardless of size, are counted and reported under this heading.
	18) Visible foreign matter: It takes both dust and trash particles as well as their size into account and expressed as percent.

Note: USTER, HVI, high-volume investigation; AFIS, advanced fiber investigation system; UHML, upper half mean length; UI, uniformity index; SFI, short fiber index; UQL, upper quartile length; Rd, reflectance.

10.9 Testing of Yarn

10.9.1 Yarn Linear density

The manufacturing and application of any kind of yarn is based on its linear density which depends upon the diameter of the yarn. Therefore, the yarn diameter is a significant factor among all the other quality parameters of yarn. The determination of yarn diameter is difficult to measure accurately. It is due to the fact that the diameter of yarn changes significantly if the yarn is compressed or stretched. Most of the yarn diameter measuring methods involve the compressing of yarn when testing is done. Due to this compressive behavior of yarn, the measured diameter varies with the pressure applied. Optical techniques of determining yarn's diameter have the problem of specifying where the peripheral edge of the yarn lies as the yarn surface can be unclear or rough due to hairiness. This attributes that the positioning of the yarn edges is subject to operator interpretation. Owing to these troubles, a system must be designed to find out the delicacy of yarn by weighing its predefined length. This quantity is called linear density which can be determined with accuracy if the tested length of yarn is sufficient. There are two systems to find out the linear density of yarn: the direct method and the indirect method.

10.9.2 Direct System of Yarn Linear Density Measurement

The direct system of linear density involves the measurement of weight per unit length of yarn. The most known direct systems in use are as follows:
– Tex = It is the number of grams in 1,000 m length of yarn.
– Decitex = It is the number of grams in 10,000 m length of yarn.
– Denier = It is the number of grams in 9,000 m length of yarn.

10.9.3 Indirect System of Yarn Linear Density Measurement

The indirect system of the linear density length per unit weight of yarn is determined. This linear density is also called count, due to the fact that it measures the number of hanks of a specified yarn length which are required to equal a fixed yarn weight. It is the most widely used system of measuring the yarn's linear density [1].

The mostly known direct systems in use are as follows:
– Yorkshire Skeins Woolen Ny
  Count = number of hanks per pound (where 1 hank = 256 yards)

– Worsted Count New
  Count = number of hanks per pound (where 1 hank = 560 yards)

– Cotton Count Nec
  Count = number of hanks per pound (where 1 hank = 840 yards)

– Metric count Nm
  Count = number of kilometers per kilogram

Note: In the indirect systems, the fineness and count yarn count are directly proportional to each other.

10.9.4 Evenness of Yarn

The final appearance of woven and knitted fabrics is highly dependent on yarn evenness. The yarn evenness is usually defined as the variation in its thickness or in weight per unit length of the yarn. The evenness of yarn is measured by the methods described further. The conversion factors are given in Table-2.

Tab. 10.2: Conversion table.

Units	Tex	Dtex (decitex)	den (denier)	Nm (number metric)	Nec (Number English cotton)	Worsted
Tex		10 dtex	9 den	1,000 Nm	591 Nec	885.8 worsted
dtex (decitex)	Tex × 10		0.9 den	10,000 Nm	5,910 Nec	8,858 worsted
den (denier)	Tex × 9	dtex × 0.9		9,000 Nm	5,319 Nec	7,972 worsted
Nm (number metric)	1,000 tex	10,000 dtex	9,000 den		Nec × 1.693	0.886 worsted
Nec (Number English cotton)	591 tex	591 dtex	5,319 den	Nm × 0.59		1.5 worsted
Worsted	885.8 tex	8,858 dtex	7,972 den	1.129 Nm	0.666 Nec	

10.9.5 Visual Examination

In this method, yarn evenness is checked by wrapping it on a blackboard in uniformly spaced turns to reduce the effect of optical illusions caused by irregularity. The prepared yarn boards are analyzed by using a uniform and unidirectional source of light. Normally, visual examination is done without any comparison with a standard, but a comparison could also be made with the ASTM standard if it is available. Nowadays more uniformly spaced yarn boards are prepared with the help of motorized wrapping machines as shown in Fig. 10.1. By these wrapping reels, the yarn moves continuously and steadily on the tapered revolving blackboard. Tapered blackboards are preferred to evaluate or determine periodic faults. If there are periodic faults in the yarn, they produce a woody pattern which is clearly visible. This visibility of the yarn faults on the tapered boards is due to the equal spacing of the yarns on the board [1].

➤ Yarn Sample

Fig. 10.1: Yarn blackboard for visual examination.

10.9.6 Cut and Weight Technique

It is the simplest and most used method of measuring variation in mass/weight per unit length of yarn. In this technique, the accurate cutting of consecutive lengths from the yarn sample is done. The weight of these cut lengths is measured by a sensitive weighing balance. For this testing, an accurate method of yarn cutting is needed to achieve the same lengths. A small error in cutting the lengths of yarn can cause an equal error in the measured weight which leads toward wrong results. For this purpose, yarn is wrapped around a grooved rod which has a circumference of exactly 2.5 cm, then a sharp blade is run along the groove which gives the yarn in equal 2.5 cm lengths. The lengths produced can be weighed on a suitable sensitive balance. The mass of each cut length of yarn is plotted on a graph as shown in Fig. 10.2.

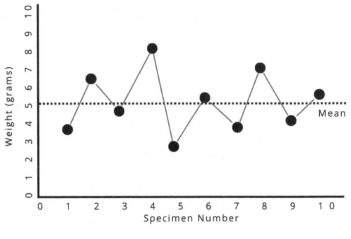

Weight of consecutive 25 mm specimen

Fig. 10.2: Variations in mass/weight per unit length of a yarn sample.

A line on the plot representing the mean value can then be drawn. The scatter of the individual values from the drawn line will provide a visual clear indication of the yarn unevenness. The mathematical measurement of yarn unevenness is required, which will take account of the distance of the individual points from the mean line and the number of them. Usually, two methods are employed to express it.

The average value for all deviations from the mean is calculated and expressed as a percentage of the overall mean which is known as the percentage mean deviation (PMD). This term is used by USTER Technologies to indicate U%.

The value of standard deviation is expressed by squaring the deviations from the mean value which presents the percentage (%) of the overall mean value.

The deviations having a normal distribution about the mean are correlated as follows [1]:

$$CV = 1.25 \times PMD$$

10.9.7 USTER Yarn Evenness Tester

The USTER yarn evenness tester finds the variations in the thickness of yarn by using capacitive techniques. The yarn to be examined is passed through a pair of parallel capacitor plates whose capacitance is continuously measured electronically. The capacitance of the system changes continuously due to the presence of yarn between the capacitor's plates. The capacitance depends on the mass of a material (yarn) between the plates and its dielectric constant (the type of raw material used). For the same dielectric constant, the signals are directly related to the mass of yarn between the capacitor plates. To get the same relative permittivity for yarn, it should be made up of the same type of fiber and it must have uniform moisture content throughout its length. The varying moisture content or an uneven blend of two or more fibers will vary the dielectric constant in different parts of the yarn, and this variation will be signaled as unevenness. The readings made by the USTER tester are equivalent to weighing successive 1 cm lengths of the yarn [3].

The following can be the possible reasons for yarn unevenness:

- The number of fibers in the yarn cross section is not constant; it varies widely depending on the fiber parameters. This is the most significant reason for yarn unevenness.
- The staple spun yarn is made up of natural fibers having variable fineness. This variation leads toward different yarn thicknesses, even when the same amount of fibers are present in the cross section of yarn.

10.9.8 Yarn Hairiness

The yarn hairiness highly affects the quality of the yarn and fabric. The existence of the lint in the yarn gives a poor appearance of woven or knitted fabric. The Uster reports showed that 15% of the fabric faults are caused by the hairiness of yarns. The yarn hairiness is the ratio of the length of all protruding fibers in 1 cm and the yarn's actual length (1 cm). It is a unitless parameter. It is an undesirable parameter which directly influences the fabric production. Therefore, it is very important to measure it so that it can be controlled. The hairiness value of the coarser yarns is normally lower than the finer yarn due to good cohesive/frictional forces between the constituent fibers in the coarser yarns. Shirley yarn hairiness tester, Zweigle G565, and Uster tester 5 hairiness meter attachment are used for testing the hairiness of the yarns [1].

10.9.9 Yarn Tensile Strength

The yarn tensile strength is the maximum force/load that is required to break or rupture the yarn along its length [4]. The strength of yarn mainly depends upon the type of fibers, level of twist, and methodology to prepare the yarn. The tensile strength of the yarn is an important parameter regarding the fabrication of the yarn because it directly influences the strength of the developed fabrics. Two different approaches are used to measure the yarn strength. In the first approach, single yarn strength is determined. Normally Newton (N) and cN units are used for the yarn strength [5]. The amount of force which is applied on the material to produce an acceleration of meter/second square per second is 1 N [1]. Single yarn strength provides information about the warping machine and loom efficiency. In order to calculate the combined strength effect of the yarn, count lea strength product of the yarn is calculated. A lea of 120 yards is made with the help of a wrapping reel, and the weight of the lea is determined in order to calculate the yarn count by using the formula number of hanks/pound [6]. Lea strength is determined by using the lea strength machine which has two jaws: one is fixed and the other is attached to the load. Using the constant rate of loading principle, the tensile strength of the lea is determined [7].

10.9.10 Yarn Twist Measurement Test

The twist provides cohesion to the staple fibers in spun yarn. The strength, appearance, and handle of yarn depend upon the level of twist inserted in the yarn. Therefore, to check the exact level of twist inside the yarn, the twist measurement is done. The twist measurement is commonly done by using a twist tester which works on the principle of removing the twist by rotating the yarn mechanically. A 25 m of the yarn sample is placed between moveable clamps of the tester, and tension of 0.25 cN/tex is maintained.

After setting the tension, the twist is removed completely by turning the rotatable clamps of the machine until fibers inside the yarn become parallel. Then the total number of rotations for untwisting of yarn provides the twist in the set length of clamps. The number of turns are counted and the turns per unit length of yarn sample are calculated.

10.10 Machines for Fabric Tensile Strength Testing

On the basis of the working principle, tensile strength testing machines can be categorized into three major categories:

1. Constant rate of extension (CRE): the machines in which the rate of elongation of the specimen is constant and with increasing load, there is a negligible movement of the load measuring mechanism. The working principle of Tensorapid-4, which is used to evaluate the tensile strength of a single yarn, is the constant rate of elongation [1].
2. Constant rate of loading: the machines in which the applied load on the test sample is increased constantly with time. In this, the specimen is free to elongate and the extension of the specimen depends upon its properties at any applied load [1]. The working principle of the lea strength machine is the constant rate of loading.
3. Constant rate of traverse: in such types of machines, two pulling clamps are used to evaluate the tensile strength of the sample. One clamp moves with a uniform rate and the load is applied by the other clamp which moves to activate a load measuring mechanism. In such machines, the rate of increasing load/elongation is usually not constant. Therefore, the rate of increase of loading or elongation is not constant which depends upon the extension properties of the yarn/fabric sample [1]. Normally, the old testing machines use the abovementioned working mechanism such as the old fabric tensile strength testing machine.

10.11 Fabric Tests

After the development of fabric, different tests are performed on it to evaluate the performance parameters by keeping an eye on the end applications of the fabric. The strength, appearance, and abrasion resistance are of great importance for any type of fabric. Different strength tests are performed to evaluate the fabric strength. The strength evaluation of fabric is done differently for woven and knitted fabric. For woven fabrics, tensile and tear strength are determined while the bursting strength is measured for knitted fabric.

10.11.1 Strip Test

In this test, five different samples are prepared which are expanded in the direction parallel to ends (warps) and five other samples are parallel to the weft yarn. The sample 60 mm × 300 mm is cut from the fabric. After cutting, the sample is frayed down from both sides in width equally to produce a fabric sample of 50 mm thickness for the final test. The fraying of the sample is done in order to minimize the human error which we have to face in case of improper cutting of the sample. The gauge length is 200 mm, and the speed of the extension is 50 mm/min. The mean breaking force and mean extension as a percentage of initial length are reported.

10.11.2 Grab Test

There are three different approaches to produce the fabric samples for the tensile strength test. These methods are given as follows:

- In this approach, a fabric sample of 25 mm × 50 mm is prepared as done in the strip method.
- The cut fabric strips of 25 and 50 mm in width are taken to be used with fabrics that cannot be easily frayed such as heavy milled fabrics.
- The grab method is relatively simpler and quicker. Jaw faces are used (Fig. 10.3) in this method, which is considerably narrower than the fabric, so fraying the fabric to width is not required. In this test, only a 25 mm sample is tested in the jaws.

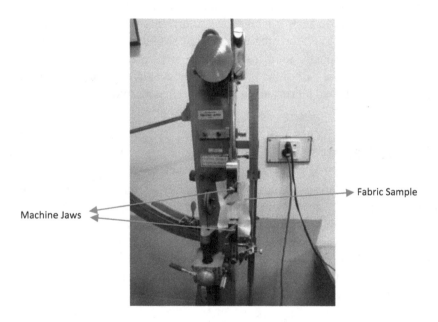

Fabric Sample

Machine Jaws

Fig. 10.3: A grab method tester for fabric strength.

10.11.3 Tear Tests

Tear strength is an important parameter to analyze the fabric performance during its exposure to the swear environmental conditions. This testing method is one of the most commonly used methods to analyze the strength failure of different kinds of fabrics. The tear strength testing is an important evaluation regarding the functional or industrial fabrics which are usually exposed to rough handling and higher loads during usage such as tents and canopies. In the case of outdoor clothing, the overall and casual dress tearing strength is an important characteristic.

Tearing strength can be defined as the force required to propagate an existing tear in the fabric. Different tear tests are performed in order to measure the tear tests. The basic principle of testing is same and only the geometry of the specimen is different. The rip tear test is the simplest test. A cut is introduced in the center of a strip of the specimen and by the help of the jaws of a tensile strength tester, two tails are pulled apart and the tear strength is evaluated [8].

10.11.4 Elmendorf Tear Tester

The pendulum-type ballistic tester is used to tear the fabric sample, and the energy loss during tearing is measured. The following formulas are used to determine the energy loss during tearing:

$$\text{Energy loss} = \text{force required to tear} \times \text{distance traveled}$$

In this case, the loss in potential energy will be equal to the work done. The apparatus used in this test is presented in Fig. 10.4. The sample is fixed between the clamps of the tester and the tear is started by a slit cut in the fabric sample which is fixed between the tester clamps. The pendulum is then released to determine the tear strength, and the specimen is torn as the moving jaw moves away from the fixed one. The potential energy of the pendulum is maximum at the starting position due to its height. As a result, tearing the energy of the pendulum is lost and it does not get the same height from

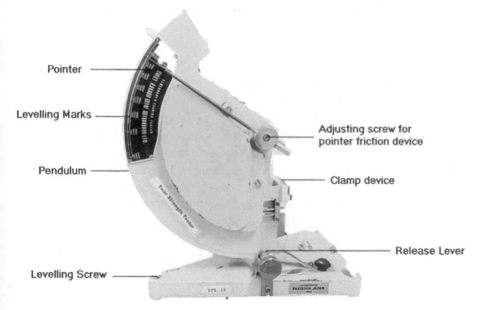

Fig. 10.4: Elmendorf tear strength tester.

which it was released. The energy lost during the specimen is directly proportional to the difference between the initial height and the final height after energy loss. By this tester, the tear strength or percentage of the original energy is determined [9].

10.11.5 Measurement of Fabric Bursting Strength

The bursting strength is usually checked for knitted fabrics. It is defined as the force required to rupture a woven/knitted fabric by dilating it with a force which is applied perpendicular to the plane of the sample under slandered conditions.

The machine for this test works on the principle of CRE. The sample is securely clamped to the machine without any kind of tension on the ball burst attachment. A force is applied against the fabric sample with the help of a polished, hardened steel ball until rupture occurs [10].

10.11.6 Weight per Square Meter (GSM) of Fabric

The mass/weight per unit area (g/m^2) of the fabric sample is determined by using the GSM cutter. In order to determine the weight per unit area of the fabric, samples of 113 mm diameter are prepared from different places with the help of a GSM cutter. Five random samples are prepared, and the weight is determined with the help of weighing balance. The sample weight is determined in grams; it is divided by the constant (0.01) in order to determine the gram per square meter. Gram/square yard can also be determined by this method [11].

10.11.7 Color Fastness to Perspiration

Color fastness can be defined as the resistance of the fabric sample to change its color properties to the adjacent materials or both on the exposure of the sample to any environment that may take place during the processing, testing, storage, and so on [12].

Perspiration is a saline fluid which is produced by the sweat glands. In order to test the color fastness of the fabric sample, two textile specimens are used. Both samples are wetted in the simulated acid perspiration solution. The test samples are wetted under controlled specific mechanical pressure. After wetting, the specimens are dried at a slowly elevated temperature. After the conditioning process, the color change of the specimen is checked and other specimens or fiber materials are checked for color transfer [12].

10.11.8 Fabric Abrasion Resistance (Accelerator Method)

10.11.8.1 Abrasion Resistance

When a mechanical action takes place between the two fabric surfaces, distortion of the samples takes place. Resistance against this wearing action is called abrasion resistance.

An unfettered fabric specimen is used for this test, which is driven in a zigzag course with the help of a rotor generally in a circular orbit in a cylindrical orbit. The specimen is subject to high velocity and rapid impacts. During movement in the cylindrical orbit, the sample undergoes different mechanical actions such as compression, shock, and stretching. Rubbing action of yarn against the yarn, surface against abradant, and surface against surface produces abrasion in the test sample as shown in Fig. 10.5. Weight loss of the specimen or strength loss of the woven specimen determines its abrasion resistance. Change in other fabric characteristics is also used to evaluate the abrasion resistance of the specimen [13].

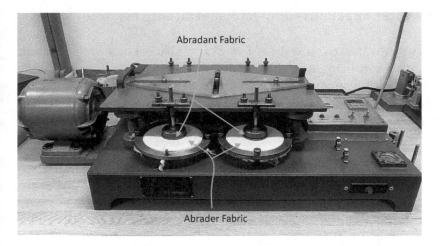

Fig. 10.5: Fabric abrasion tester.

10.11.9 Water Repellency (Spray Test)

The ability of any material to resist wetting against water is called its water repellency. The water resistance test of industrial or functional fabrics has great importance. In this test, water is allowed to spray (Fig. 10.6) on the stretched sample of fabric from 150 ± 2 mm height under controlled conditions. A wet pattern is produced whose size is dependent on the water repellency of the specimen. The wet fabric sample is compared by the standard spray ratings to determine its water repellency [5].

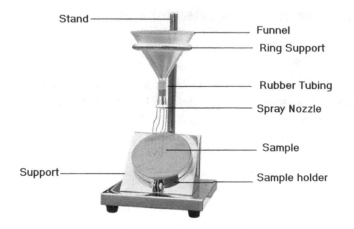

Fig. 10.6: Spray tester.

10.11.10 Water Repellency/Impact Penetration Test

In this test, water is sprayed on stretched fabric sample backed by a weighted blotter under controlled conditions. After the spraying, the blotted paper is reweighted in order to determine the water penetration. If the weight difference of the blotted paper is more, then the water repellency of the fabric will be less and vice versa. The diagram of the impact penetration apparatus is given in Fig. 10.7 [14].

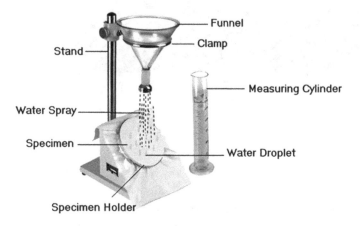

Fig. 10.7: Impact penetration.

10.11.11 Air Permeability

Air permeability of any material can be defined as the air passing at a right angle through a known area of the specimen in a unit time under a prescribed air pressure differential between the two surfaces of a specimen.

In order to evaluate the air permeability of the specimen, air passing perpendicularly through the specific area of the fabric in a unit time is adjusted in order to obtain the predetermined air pressure difference between the two surfaces of the samples. This rate of airflow helps to determine the air permeability of the fabric sample. This parameter directly influences the comfort parameter of the fabric [15].

10.11.12 Moisture Management Test

This test is performed to determine the dynamic liquid moisture transport behavior of the sample in multiple directions. A moisture management test is performed by placing the samples between the two horizontal sensors. Each horizontal sensor has seven concentric pins. In order to measure the electrical conductivity changes, a specific volume of the test solution is dropped onto the center of the specimen facing upward. The test solution is free to move in multidirections. Changes in the electrical resistance of the sample during the test are determined and recorded [16].

10.11.13 Color Fastness

It is the resistance of the material to its change in the color properties, transferring its colorants to the other materials.

Crocking is the transfer of the colorants from the surface of the colored fabric or yarn to an adjacent area of the same fabric or the other fabric principally by rubbing. Under controlled conditions, the colored specimen is rubbed with the white test cloth. Color transfer to the white cloth is evaluated by comparison with the standard gray scale [17].

10.11.14 Thermal Transmittance of Textile Materials

It can be defined as the time rate of unidirectional heat transfer per unit area in the steady state between parallel planes per unit difference of temperature of the planes.

For this test, a dry fabric sample is used. In this test, the overall thermal transmission coefficient of the specimen is determined as a result of the combined action of the convection, conduction, and radiation for the dry specimen and other materials within the limits. The unit of the thermal transmittance of the textile materials is $W/m^2 \cdot K$. The permeability of air, breathability, and thermal conductivity of the fabric directly influence the comfort parameters of the fabric samples [18].

10.11.15 Effect of Repeated Home Laundering on Textiles

Laundering of the textile materials is a process which is performed to remove the soil and stains of the fabric by treating it with an aqueous detergent solution. Rinsing, extracting, and drying processes are performed for laundering. Repeated home laundering effect on the appearance of the specimen is evaluated by this test method [19].

10.11.16 Color Fastness to Perspiration

This test method is used to evaluate the resistance against the color change of the colored textiles as a result of perspiration. Alkaline and saline solutions which contain histidine are used to treat the composite specimens, drained and placed in a Perspirometer for 30 min at room temperature. The colored fabric sample and the undyed cloths are removed from the solutions and dried separately. The staining of the undyed cloths and the change in color of the specimens are evaluated by comparing with the standard gray scales [1].

10.11.17 Seam strength

The joining line of two or more fabrics is called a seam. Seam failure of the garment can be defined as the point at which sewing threads or the fabric ruptures or causes excessive yarn slippage adjacent to the stitches. Seam strength is an important parameter which should be considered during garment making. There are different possible causes of seam failure:
- Wear out or fail of the sewing thread before the fabric
- Damaging of the sewing thread during the sewing process by the needle
- Slippage of the seam takes place

In order to evaluate the seam slippage of the fabric, two samples, one with seam and one without seam, are prepared from the fabric having the same warp and weft density. The sample without a seam is stretched first with the help of the tensile tester, the applied load limit is up to a load of 200 N, and a force elongation curve is drawn. By using the same method, fabric sample having a seam is tested. Both curves are compared. Horizontal separation of the curves will be due to seam slippage.

References

[1]   B. P. Saville, *Physical Testing of Textiles*. Cambridge, England: Woodhead Publishing Limited, 1999.
[2]   BS EN ISO 139:2005, "Textiles-Standard atmospheres for conditioning and testing," 2005.
[3]   Standard Test Method D 1425–96, "Standard Test Method for Unevenness of Textile Strands Using Capacitance Testing," 1996.
[4]   Designation: D 123–01, "Standard Terminology Relating to Textiles," 2002.
[5]   ASTM D 2256–97, "Standard Test Method for Tensile Properties of Yarns by the Single-Strand Method," 1997.
[6]   Designation: D 1907–01, "Standard Test Method for Linear Density of Yarn (Yarn Number) by the Skein Method," 1997.
[7]   Designation: D 1578–93, "Standard Test Method for Breaking Strength of Yarn in Skein Form," 2000.
[8]   Designation: D 2256–96, "Standard Test Method for Tearing Strength of Fabrics by the Tongue (Single Rip) Procedure (Constant-Rate-of-Extension Tensile Testing Machine)," 1996.
[9]   Designation: D 1424–96, "Standard Test Method for Tearing Strength of Fabrics by Fallingyoe-Pendulum Type (Elmendorf) Apparatus," 1996.
[10] Designation: D 6797–02, "Standard Test Method for Bursting Strength of Fabrics Constant-Rate-of-Extension (CRE) Ball Burst Test 1," 2002.
[11] ASTM 3512, "Standard Test Method for Pilling Resistance and Other Related Surface Changes of Textile Fabrics: Random Tumble Pilling Tester,"2014.
[12] AATCC Test Method 15–2002, "Colorfastness to Perspiration,"2002.
[13] AATCC Test Method 93–1999, "Abrasion Resistance of Fabrics: Accelerator Method," 1999.
[14] AATCC Test Method 42–2000, "Water Resistance: Impact Penetration Test," 2000.
[15] Designation: D 737–96, "Standard Test Method for Air Permeability of Textile Fabrics," vol. 14, pp. 1–5, 1996.
[16] AATCC Test Method 195–2011, "Liquid Moisture Management Properties of Textile Fabrics," 2011.
[17] AATCC Test Method 8–2001, "Colorfastness to Crocking : AATCC Crockmeter Method,"2001.
[18] ASTM D 1518–85, "Standard Test Method for Thermal Transmittance of Textile Materials," 1998.
[19] AATCC Test Method 124–2001, "Appearance of fabric after repeated home laundering," 2001.

Sajjad Ahmad Baig

11 Quality Control and Quality Assurance

Abstract: There can be no room for compromise when it comes to quality in any industry. In today's economy, customers demand and expect a return on their investment. There must be a constant effort to produce high-quality work as a clothing manufacturer. Industrialists or manufacturers must manage and monitor the quality of products and services through quality control (QC) and quality assurance (QA) systems. These are the key factors for the increased competitiveness of their companies. QC and QA systems together constitute the key quality systems and the parts of quality management. QC is concerned with ensuring that quality requirements are met, and assurance is concerned with ensuring that they will be met. The quality systems must be commensurate with the company's business objectives and business model. Top management commitment and active involvement are critical for the successful implementation of quality systems.

Keywords: Quality control, quality assurance, product development

11.1 Introduction

All the individuals in this sphere know about the term "quality" and expect the services and products he is buying to be of the best quality. The concept and vocabulary of quality are indefinable. For different people the concept of quality is different. Often, customers will say, "I only ever buy my clothes from Uniworth because their shirt size 15 fits my figure perfectly." as an example of how a customer perceives quality. "I buy all of our family's clothes at Wal-Mart because they don't need ironing when tumble-dried." Many people prefer a brand that celebrities wear, and this is their definition of quality. A small number of persons can explain quality in quantifiable terms that can be operationalized. When asked what differentiates their product or service, the banker will answer "service," the healthcare worker will answer "quality health care," the hotel restaurant employee will answer "customer satisfaction," and the manufacturer will simply answer "quality product." There is an old maxim in management that says, "If you can't measure it, you can't manage it," and so it is with quality. If strategic management systems and competitive advantage are to be based on quality, every member of the organization should be clear about this concept, definition, and measurement as it applies to his or her job.

Harvard professor David Garvin, in his book *Managing Quality*, summarized five principal approaches to defining quality: transcendent, product-based, user based, manufacturing based, and value based [7]. Let us discuss each one of them.

https://doi.org/10.1515/9783110799415-011

11.1.1 Transcendental View of Quality

Those who hold a transcendental view would say, "I can't define it, but I know when I see it." Advertisers are fond of promoting products in these terms. "Where shopping is a pleasure" (supermarket), "We love to fly and it shows" (airline), and "It means beautiful eyes" (cosmetics) are examples.

11.1.2 Product-Based View

Product-based definitions are different. Quality is viewed as quantifiable and measurable characteristics or attributes. For example, durability or reliability can be measured (e.g., mean time between failure, fit, and finish), and the engineer can design to that benchmark. Quality is determined objectively. Although this approach has many benefits, it has limitations as well. Where quality is based on individual taste or preference, the benchmark for measurement may be misleading.

11.1.3 User-Based View

User-based definitions are based on the idea that quality is an individual matter, and products that best satisfy their preferences (i.e., perceived quality) are those with the highest quality. This is a rational approach but leads to two problems. First, consumer preferences vary widely, and it is difficult to aggregate these preferences into products with wide appeal. This leads to the choice between a niche strategy and a market aggregation approach which tries to identify the product attributes that meet the needs of the largest number of consumers.

11.1.4 Manufacturing-Based View

Manufacturing-based definitions are concerned primarily with engineering and manufacturing practices and use the universal definition of "conformance to requirements." Requirements or specifications are established design, and any deviation implies a reduction in quality. The concept applies to services as well as products. Excellence in quality is not necessarily in the eye of the beholder but rather in the standards set by the organization.

This approach has serious weaknesses. The consumer's perception of quality is equated with conformance and hence is internally focused. Emphasis on reliability in design and manufacturing tends to address cost reduction as the objective, and cost reduction is perceived in a limited way – invest in design and manufacturing improvement until these incremental costs equal the costs of nonquality such as rework or scrap.

11.1.5 Value-Based View

Value-based quality is defined in terms of costs and prices as well as a number of other attributes. Thus, the consumer's purchase decision is based on quality (however it is defined) at the acceptable price. Mainly, industries are categorized into two types: product manufacturing and service providing. Nowadays the textile industry comes under the product manufacturing category; a lot of resources are being utilized yearly throughout the world to develop and introduced a system to produce reliable quality products and services. International competition among producers is diverting manufacturers from quantity production to highly reliable and quality products and services. Now, the big question is "what makes the quality?" When the same technology is adopted at different textile mills, do they get the same quality all the time? Why there is a difference in the quality of the products of both mills? Even after spending billions on buying the latest machinery and best raw materials and hiring the best people, manufacturing companies do not achieve the best quality. The correct answer to these questions is the difference in the work quality of both mills. The highly motivated human resource of the company is more important to get the best quality. To drive the workers to achieve quality should be emphasized more than anything else. When the management succeeds in involving the people at work, the quality can be achieved even with old machines; of course, the productivity may be less. What additional work do they do when working by heart? Do they adopt any different settings or speeds? No, it is the same machine, the same setting, but when the work is done devotedly, the mistakes are not there. Only good quality is produced. According to the gurus of total quality management (TQM), quality is a firm's strongest weapon that can take an organization higher and higher even among competitors. Simply quality means customer needs to be satisfied. Failure to maintain a sufficient and suitable quality standard can be an unsuccessful and challenging thing for that firm. But maintaining an adequate standard of quality also costs effort. From the first investigation find out what the potential customer for a new product wants, through the processes of design, specification, controlled manufacture, and sale.

Lots of firms are working in the global village but their production and checking or controlling system vary from firm to firm. Quality has a significant role in the industrial revolution, and the development of work norms, being the oldest among industries, the textile and garment industry has taken the development of a number of statutory, legal, and regulatory requirements, development of new management techniques, and development of norms for industrial relations and so on. In spite of the industry being the oldest and having undergone various ups and downs, even today it is not in a position to stabilize itself and be a role model for other industries. The problems faced by the industry and the employees rather than getting solved are getting increased. Developments of technology, automation, and computer-aided techniques have helped the industry in getting productivity and quality, but the same is not getting sustained. Customers are able to tell precisely the quality they require, and in

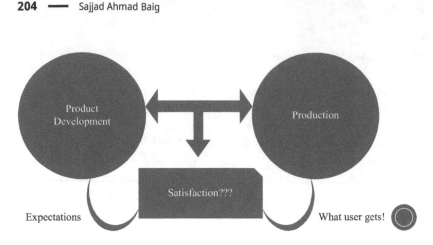

Fig. 11.1: Value-based view.

the fashion world, all the earlier so-called mistakes or poor quality are getting a different name as highly fashionable. The people do not prefer to work in the textile or garment industry due to various reasons, and the management is not trying to retain the people who are interested in working in the industry. The textile and garment industry, which was once the backbone of advanced countries, has lost its base and has shifted to developing countries. India being the cradle of civilization and mother of textiles naturally has an edge, but still, the industry is not doing well. The management is blaming staff and workers, and the employees are blaming management. The managements want the workers to give more efficiency but are overloading them with continuous work, unstable administration, and are respecting them as a part of their industry. All are trying to achieve quality and productivity by installing the latest technology and paying huge salaries to the top persons in the organization but are not addressing the basic requirements of clean administration, improving the quality of work, developing harmony among the staff and workers, and bringing a feeling of oneness among all. The people with power are carried away by the short-term plans as it looks lucrative and are not making any efforts to make the base stronger as shown in Fig. 11.1.

There are two quality systems, first one is the "quality control" (QC) and the second one is "quality assurance" (QA). The manager of a firm and also the culture of the firm used to decide which one is their primary tool to provide quality products to their customers. QC is a reactive approach and on the other end, QA is a proactive approach. QA and QC are complicated concepts for the textile industry. QA is not the same as QC, but QC is a component of QA. QA is defined as the "process of designing, manufacturing, testing, and evaluating products to ensure that they meet the desired quality level for a company's target market" ([14], p. 6). QA examines a product from its inception until it is sold to the consumer. QC is commonly understood as assessing quality after products have been manufactured and categorizing them as acceptable or unacceptable.

With that approach, we can only assure you that we have checked all the production and removed all the defective ones. But it is not going to take us higher in quality manufacturing, we have to control and take care of our production through improvements in the process. When we are improving our system to avoid and stop defects, this is a QA approach; but if we just remove defects after production, then it will be a QC approach.

11.2 Quality Control

QC is confidently the oldest system and its roots sink into the initial statistical research carried out by Shewart [4]. These principles were further developed in Japan after the end of the Second World War. Feigenbaum developed total QC, defining it as (1961) "a network of the management/control and procedure. That is required to produce and deliver a product with a specific quality standard." QC is a reactive approach. Whenever products or services do not satisfy requirements set by customers or by the organization itself, a nonconformity is generated, with its related costs of poor quality. "The systems required for programming and coordinating the efforts of the various groups in an organization to maintain the requisite quality." QC is seen as the agent of QA or total QC. QC is a set of activities that ensures the quality of products. It focuses on identifying defects in the actual products produced. QC aims to identify and correct the defects in the finished product. The goal of QC is to identify the defects after a product is developed and before it is released.

However, in modern times the Western Electric Company was the first to use basic QC principles in design, manufacturing, and installation. In 1916, C. N. Frazee of Telephone Laboratories successfully applied statistical approaches to inspection-related problems, and in 1917, G. S. Radford coined the term "quality control" [1, 6]. In 1924, Walter A. Shewhart of Western Electric Company developed QC charts. More specifically, he wrote a memorandum on May 16, 1924, which contained a sketch of a modern QC chart. Seven years later, in 1931, he published a book entitled *Economic Control of Quality of Manufactured Product* [2].

The process of range building, sourcing, planning, and controlling production varies from company to company, and the responsibility for QC also depends very much on how the company operates. It is fair to say that quality is everyone's responsibility; however, individuals with specific responsibilities have to be appointed. There are several titles used in the industry to describe this role: QA manager; quality controller; technical manager; product development manager; technical designer; and garment technologist. The role can be divided into two main areas of responsibility: technical and quality auditing. (1) Technical responsibilities include patterns, size charts, specifications, fabric approval, organizing fittings, and keeping up-to-date with new technology. (2) Quality audit's responsibilities include assessment of factories, dealing with quality issues, analyzing why faults are happening, and checking that the production is correct.

So, technology is in the continuous process of development to produce the best quality products. There are a number of factors on which the quality fitness of the textile industry is based such as – performance, reliability, durability, visual and perceived quality of the yarn, garment, finishing, designing, and clothing. Quality needs to be defined in terms of a particular framework of cost. QC departments play a pivotal role in producing good quality textile products in a textile organization/mill/factory. They perform functions such as the one shown in Fig. 11.2 [13]. The history of QC in the textile industry may be traced back to Zhou Dynasty (eleventh to eighth century B.C.) in China. For example, one dynasty decree stated that "Cotton and silks of which the quality and size are not up to the standards are not allowed to be sold on the market" [3, 11]. In the modern context, the first application of statistical QC concepts appeared to be in yarn manufacturing products during the late 1940s and 1950s [2]. In 1981, one of the largest textile companies in the world, Milliken & Company, launched its TQM efforts specifically directed to make a commitment to customer satisfaction pervading all company levels and locations. By 1989, it was ahead of its competition with respect to all measures of customer satisfaction in the United States and won the Baldridge Quality Award. According to Dhillon [6], there were around 30,000 textile-related companies in the United States, and many of them have implemented quality management initiatives for reducing costs and improving both products and customer satisfaction.

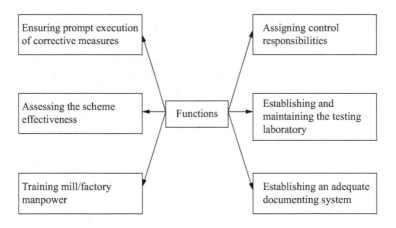

Fig. 11.2: Quality control department function.

Assigning control responsibilities is concerned with defining and assigning control responsibilities for items such as checking, controlling measurements, and weighing waste throughout the factory/mill. Training mill/factory manpower is concerned with planning and providing appropriate quality-related training to factory/mill personnel. Establishing and maintaining the testing laboratory is concerned with setting up and maintaining the testing laboratory with appropriate equipment and qualified man-power. Ensuring prompt execution of corrective measures is concerned with coordinating corrective actions in such a manner that take minimum time between the discovery of faulty operation and corrective measure. Assessing the scheme's effectiveness is concerned with regularly reviewing the scheme and making changes as considered appropriate. Establishing an adequate documenting system is concerned with designing forms for purposes such as recording measurements, calculations, summaries of measurement changes with time, and control charts.

For example, in the garment industry, QC is practiced right from the initial stage of sourcing raw materials to the stage of the final finished garment. In the textile industry, quality control is practiced from the initial stage of sourcing raw materials to the stage of final finished garment and clothing. For the textile and apparel industry, product quality is calculated in terms of quality and standard of fibers, yarns, fabric construction, color fastness, surface designs, and the final finished garment products. However, quality expectations for export are related to the type of customer segments and the retail outlets.

There are two types of QC:
- Online quality control system
- Offline quality control system

11.2.1 Online Quality Control System

This type of QC is carried out without stopping the production process. During the running of the production process, a setup is automatically performed and detects the fault and also takes corrective action. Online QC contains the raw material QC and process control and all other precautions related to manufacturing when the product is in line for assembly, it also needs to be taken care of in all the processes to make a fine finished product which is included in the online QC system.

11.2.2 Offline Quality Control System

This system was performed in the laboratory and other production areas by stopping the production process consisting of fabric inspection and laboratory and other tests. Correction steps are taken according to the test result. It also needs to be a satisfying physical test and some chemical tests that ensure the producer that the product is fine.

All the statistical, scientific, and other checking methods are used to verify the quality of produced goods. A famous saying is "Quality is made in the BOD or top management's offices, not only by employee's end." There is a need to justify all the working efforts of the employees rather than using their power for only output.

Making employees an important part of the management means lower-level or front-line staff should be well aware of all of the company's doings, and there should be proper and fast communication between top and lower-level management.

If top management does so, they will be more dedicated and they will show their loyalty toward the best makers of the products. They will give you more than you expect from them. By all means, no matter how much you work or struggle for a QC system, it will never satisfy you for a long time. Every time you have to repeat all the steps of online and off-line QC system, but if you adopt the TQM approach regarding QA, you will get the benefit now and forever.

11.3 Quality Assurance

QA is a proactive approach. It is the uplifting process for the fine production and quality of production. In QA, we do not need to inspect the workers because our focus is process-oriented rather than the people. For example, Dr. Deming [9] emphasized, "Productivity increases with the improvement of quality control and quality assurance." Low quality means high cost and loss of competitive position. Companies having better quality management practices possessed better business results, including higher market share growth, and higher profitability. QA techniques, including Jidoka, Poka-Yoke, and Lean Six Sigma. Poka-yoke, as introduced by Shingo [10], are implementing simple low-cost mistake-proofing devices that detect abnormal situations before they occur or once they occur to stop production to prevent defects. The Lean Six Sigma approach, as Taghizadegan [12] explains, is a data-driven approach to find the root cause of problems and uses the define, measure, analyze, improve, maintain, and control process to organize operating processes.

Quality costs are focused on by management in pursuit of improvement in quality, customer satisfaction, increased market share, and profit enhancement. The main purpose of quality cost considerations is to warn against oncoming dangerous financial situations to the companies. Juran [8], in his famous QC handbook, used the analogy of "Gold in the Mine" which means that losses due to avoidable mistakes/defects equal the cost of quality.

The cost of quality is generally classified into four categories [5]:
1. External failure cost
2. Internal failure cost
3. Inspection (appraisal) cost
4. Prevention cost

11.3.1 External Failure Cost

The cost associated with defects is found after the customer receives the product or service. Example: Processing customer complaints, customer returns, warranty claims, and product recalls.

11.3.2 Internal Failure Cost

The cost is associated with defects found before the customer receives the product or service. Example: scrap, rework, reinspection, retesting, material review, and material downgrades.

11.3.3 Inspection (Appraisal) Cost

The cost incurred to determine the degree of conformance to quality requirements (measuring, evaluating, or auditing). Example: inspection, testing, process or service audits, calibration of measuring, and test equipment.

11.3.4 Prevention Cost

Prevention cost is the cost incurred to prevent (keep failure and appraisal cost to a minimum) poor quality. Example: new product review, quality planning, supplier surveys, process reviews, quality improvement teams, education, and training.

The first three categories of cost of quality come under QC, which deals with the cost of not creating a quality product or service, while preventive costs deal with QA.

The proponent of TQM and QA approach like Deming, Crosby, and Juran advocated that the implementation of quality management philosophy helps companies to achieve higher levels of quality and performance. For example, Deming [9] asserts, "Productivity increases with the improvement of quality. Low quality means high cost and loss of competitive position." We need to take necessary actions and corrections to make that process fit for the required quality of a thing or product. It starts from the purchasing of raw materials or required materials for production. In the beginning, purchaser of a firm needs to be very careful. Instead of saving the cost, the firm should consider quality first. If the manager can manage cost without sacrificing quality, then it is the plus point and most important thing to that firm to attain an efficiency of the resources. Later, when moving toward production ensures that the required standards are achieved, and investigation and action are taken on substandard performance.

QA is a technique for preventing errors or flaws in manufactured goods as well as problems while providing consumers with solutions or services. QA, according to

ISO 9000, is "a component of quality management focused on fostering confidence that quality criteria will be met." Thus, it is slightly different from QC.

QA is used in the preproduction stage of physical items or services to make sure that what will be produced complies with customer specifications and requirements. QA is also applied to software to verify that features and functionality meet business objectives, and that code is relatively bug-free prior to shipping or releasing new software products and versions.

To ensure that requirements and objectives for a product, service or activity are met, administrative and procedural tasks are put into place in a quality system. This is known as QA. Assurance of quality is a cornerstone of the textile business. In order to prevent errors, it is necessary to measure, compare, and monitor processes on a regular basis. QC, on the other hand, is concerned with the result of the process. "Fit for purpose" and "Right the first time" are two of the QA principles that must be adhered to in order for a product to meet its intended use. Managing raw materials, assemblies, products, and components, as well as production-related services, is part of QA. For industry, as a whole, costs of quality have been estimated at between 4% and 15%. QA is used in the preproduction stage of physical items or services to make sure that what will be produced complies with customer specifications and requirements. Before selling or releasing new software products and versions, QA is also used to check that features and functionality satisfy business objectives and that the code is largely bug-free. QA is used in the preproduction stage of physical items or services to make sure that what will be produced complies with customer specifications and requirements before selling or releasing new software products and versions.

Successful quality management requires an orientation toward quality that permeates the entire organization. Many physical and emotional, as well as physical and mechanical, factors contribute to the production of high-quality consumer goods. A top executive must establish quality management as an ongoing part of the organization and provide the equipment, supplies, personal, and budget to support its existence. Continual improvement in quality comes only with the commitment of managers and employees to consistency, and high-quality performance in all aspects of the business.

Firms that have adopted this total concept of quality often use the terms quality assurance and total quality management. The acronym QA is used throughout this discussion. Under a QA system, evaluation of conformance to standards involves the performance of all the company's divisions as well as the products and services that are produced by the firm. It is recognized that the production of quality products depends on the quality consciousness of the entire organization, including merchandising, marketing, finance, operations, and production.

An integral aspect of any QA plan involves including employees from all levels of the organization. Employees are taught how to identify the root causes of product faults and how to fix them so that they do not happen again in the future. Only a small percentage of rejected clothing has faults that may be traced back to a sewing operator, according to certain studies. To ensure that items are manufactured in accordance with requirements, QC, a more restricted version of quality management, is used. Instead of looking at quality as a company-wide responsibility, QC activates tend to focus just on the production process.

11.3.5 Quality Assurance Policies and Records

The QA program's components are frequently divided into three levels, which are variously labeled:
(1) the strategic level (quality manual);
(2) the tactical (concerning general practices such as training, facilities, and QA operation);
(3) the operational level (dealing with the standard operating procedure (SOP) worksheets and other aspects of day-to-day operations).

11.3.6 Setting Up the System

QA can be established by adopting various methods. Each organization has its own problems and requires special consideration and planning. However, once the decision to implement a QA system has been taken and the essential funds and facilities have been made available, then a plan must be drawn up. For a new project, the QA system can be drawn up before the start but if the project is already established then a QA system can be retrofitted. In the latter situation, existing practices must be evaluated with respect to QA needs and any QA checks and procedures that are already in place. It is better to build on procedures already in place and only to remove them if they are clearly inadequate. If too many changes are enforced too quickly, especially where they are seen to increase workload, they are unlikely to be met with a favorable response and execution will be poor. The QA program must be seen to be practical and realistic and not include trivial or unreasonably time-consuming or difficult tasks.

To begin any QA program, the initial step is to create a quality policy statement so that the complete QA program can have a clear and identified road map of the process and steps. A quality policy statement, including objectives and commitments, consists of the following:
– Procedures for control and maintenance of documentation
– Procedures for ensuring traceability of all paperwork, data, and reports
– The laboratory's scope for calibrations and tests
    – Arrangements for ensuring that all new projects are reviewed to ensure that there are adequate resources to manage them properly
    – Reference to the calibration, verification, and testing procedures used

11.3.7 Quality Manual

Every factory has its own policy to assure quality to get a better product. Everything is to be written in a document; how to get the best quality product and how to control quality. For QC, this document is known as a quality manual, which is followed by each and every company in the apparel industry. A complete QC manual of the garment factory actually increases the effectiveness of QC by providing guidelines to the related parties.

11.3.8 Training

The development of the program must include all staff. Typically, the management commits resources, establishes policy and standards, approves plans, assigns responsibilities, and maintains accountability. The supervisory staff takes responsibility for the development and implementation of the program and operating personnel to provide technical expertise and advice. At all stages, the operating personnel must be consulted about the practicalities of any proposed changes. In turn, they must notify management of any problems or changes that may affect the program.

11.3.9 Standard Operating Procedures

SOPs are written descriptions of all the techniques and procedures, like the use and calibration of equipment, sampling, analysis, generation of reports, data interpretation, and transportation. These are the internal reference manual for the adopted procedure and detail every relevant step. Anybody of the appropriate training level should be able to follow the SOP. These should where necessary, cross-reference other SOPs and refer to them by number. Method SOPs may originate from organizations such as the International Organization for Standardization (ISO), British Standards Institute, American Standard Technical Method, or from the instructions that come with the test kit where a commercially produced method is used.

11.4 Need for Quality Assurance in Textile Industry

The evolution and checking of the product like design, styles, and colors, suitability of components, and fitness of product is the basis of QA. The products are examined in a systematic manner to identify any errors that may be present. If these are identified then the product is said to have failed the QA, but after examination, if yarn, fabric, finish, color, and fineness meet the requirement then it is considered to be good. Ultimately it improves the quality and standard of the product.

11.4.1 Main Objectives

1. To examine the product in a systematic and comprehensive way
2. To identify areas in which standards have not been met and correct them

11.4.2 Major Responsibilities

Conduct quality process audits at manufacturing plants, as well as select suppliers, to ensure all auditing procedures are being followed. Identify quality problems and potential problems, and investigate the root cause, as well as initiate effective corrective actions in a resourceful and timely manner. Communicate effectively with others on QA activities and decisions.

The primary purpose of QA is to have feedback on the product like whether the color, texture, finishing, design, and style of the fabric are good and according to the requirement or not. Also, make sure that there is continuous development and efficiency, and effectiveness in the product.

11.4.3 Stimulate Research Effort

Research efforts are usually made to learn more about the relationship between inventories and expected outcomes. This is very important to improve the product. Other than improvement, there are new products that can be made with innovative research, like any changing needs of designing, and color schemes. This can only be possible by research in assurance.

11.5 Identify and Address the Various Problems

The various problems in a product are all identified and improved, and various procedures and outcomes are done. This is another important purpose of QA. The problem may be in the working of the product or the structure of the product itself. As if there is a defect in the finishing procedure, all is controlled easily by assurance. Similarly, all other defects in the making of fabric can be identified and addressed regularly.

11.6 Future Trends in Quality Control and Quality Assurance

What a thrilling time it is for the textile industry, with new technologies constantly dominating the market. Examining new horizons in nanotechnology textiles, conducting green or environmentally friendly textile production, sustainability and organic production are just a few examples of current and emerging trends in QC and QA in the apparel sector. Using ecologically friendly technique, or the "green" movement, has been more popular recently even though it is not a new trend. One of the major offenders and supporters of this movement is the clothing business. The entire manufacturing process used to produce garments, from the fiber to the garment, is not environmentally friendly. Denim jean manufacturing processes are being replaced with more environmentally friendly ones, according to Cotton Incorporation. The use of enzymes instead of sodium hypochlorite for bleaching and the substitution of laser etching for abrasive chemicals used in sanding and spraying are just a few examples of these techniques. A common challenge for all the exporters was the complex situation regarding export-related compliances, both at the national and international levels. Furthermore, they expressed their need for universal best practices enabling them to comply with various legal frameworks when exporting to different countries. They must comply with different Textile Organization standards such as the Better Cotton Initiative. Better Cotton Standard System, Business Social Compliance Initiative, BLUESIGN Solutions and Services, EU Ecolabel, Global Recycled Standard, OEKO-TEX, and Worldwide Responsible Accredited Production. Large textile product importers frequently establish their own codes of conduct for the exporters and manufacturers in developing countries to ensure that all standards are being followed, mindful of their brand image. The fore-mentioned firms simply do not rely upon ISO or other certifications and have their own auditors who are sent to evaluate a company's performance, for an example, IKEA buys textile products from several textile companies in Pakistan and does not permit usage of chemicals like azo dyes or cadmium to be used in textile products; therefore, it applies the strictest possible (German) regulations on pentachlorophenol to be used as a molding agent.

References

[1]   Y. M. Chae, H. S. Kim, K. C. Tark, H. J. Park, and S. H. Ho, Analysis of healthcare qualityindicator using data mining and decision support system, Expert System with Application, vol. 24, no. 2, pp. 167–172, 2003.

[2]   E. Chaplin, "Customerer driven healthcare comprehensive quality function deployment". in *56th Annual Quality Congress*. Quality Congress, pp. 767–781, 2002.

[3]   T. G. Clapp, A. B. Godfrey, D. Greenson, R. H. Jonson, and C. Seastrunk, *Quality Initiatives Reshape the Textile Industry*. 2003, Retrieved from http://www.qualitydigest.com/oct-01/html/textile.html

[4]   W. A. Shewhart, *Statistical Methods from the Viewpoint of Quality Control*. Graduate School, Department of Agriculture, Washington DC, 1939, 75.

[5]   F. A. Buttle, and M. Ross Jayne, "ISO 9000: is the real estate sector any different?", Property Management, vol. 17, no. 2, pp. 125–138, 1999. https://doi.org/10.1108/02637479910263173.

[6]   B. S. Dhillon, *Applied Reliablity and Quality: Fundamentals, Methods and Procedures*. Springer Science & Business Media, 2007.

[7]   D. A. Gravin, *Managing Quality: The Strategic and Competitive Edge*. New York: Free Press, 1988.

[8]   J. Juran, *Quality Control Handbook*. London: McGraw-Hill, 1974.

[9]   W. E. Deming, *Out of crisis, Centre for Advanced Engineering Study*. Massachusetts Institute of Technology, Cambridge, MA, 1986, 367–388.

[10]   S. Shingo, *Zero Quality Control: Source Inspection and Poka-Yoke System*. Cambridge, MA: Productivity Press, 1985.

[11]   P. Siekman, *The Big Myth About U.S. Manufacturing*. Fortune Magzine, 2000.

[12]   S. Taghizadegan, *Essentials of Lean Six Sigma*. UK: Oxford, Elsevier Inc, 2006.

[13]   United nation Industrial Development Organization, *Quality Control in Textile Industry*. New York: United Nation, 1972.

[14]   S. J. Kadolph, *Quality Assurance for Textiles and Apparel*. Fairchild Books, 2007.

Muhammad Zohaib Fazal, Syed Talha Ali Hamdani

12 Computer Applications in Textiles

Abstract: In every field, there is an implementation of computer applications. Computers have changed the sense of working and business. Bringing automation to the textile sector, changing the old tedious working techniques to the state of the art, increase in production, profits, innovation in products, a transformation from conventional to technical textiles, keeping track of the entire production process from fiber to final garment, and keeping eye on the changing marketing to survive in the competitive business environment, all became easy due to the computer applications in textiles. Not only in industry but also for research purposes, computer applications have revolutionized the possibility of novel ideas and products. This chapter is written with key examples from the weaving industry; however, similar examples are available for other textile industries too.

Keywords: computer, ERP, machine learning, image analysis, data mining

12.1 Introduction

As scientific research is progressing, human beings are being equipped with very sophisticated technology. Textile research is not only focused on clothing but also plays a vital role in other areas of research such as, electrical and electronic engineering, geo-engineering, structural engineering, composite technology, and many more. The progressive innovations in computer technologies have accelerated this research. MATLAB®, Minitab®, SPSS®, and Microsoft Office® are frequently used for statistical analysis, mathematical modeling, algorithm, and report writing. Abaqus®, COMSOL®, and Ansys® are used for finite element analysis, mechanical and electrical properties analysis, and product development. Information technology is the industry that has an impact on each and every other field of life, bringing a lot of innovations. Also, converting the old, time-consuming, manual approaches to very efficient and novel methods.

The textile industry is mainly dealing with the design and production of yarn, cloth, and garments. With the advent of human civilization, textile products (yarn and fabric) have made their significance.

The textile industry is a well-established industry all over the world, particularly in Asia, being developed to modern standards in twenty-first century with the invention of automated machines. As per data available at The Observatory of Economic Complexity (OEC), the textile makes a global trade of \$774B, with 4.62% of the total world trade, standing at seventh position of world's most traded product, in 2020. This sector demands to be equipped with modernized techniques and tools. Like other industries, the textile sector is also adding in the huge dataset generations,

https://doi.org/10.1515/9783110799415-012

needs to be analyzed for useful knowledge. It will ultimately lead to the betterment and progress of this sector with improved and enhanced techniques. The inclusion of computer technologies techniques will result in an increase in profitability and production of the textile industry.

In this chapter, we cover the aspects of the textile sector, getting benefits from computer applications and information technology techniques.

12.2 Industrial Applications

In the textile industry, the induction of computers and their applications has revolutionized manufacturing techniques. Product development is more innovative than ever. In addition to this, the time consumption in generating an idea and then its conversion to the end product has also been reduced to a greater extent. There are several other advantages that the manufacturing units are getting. Some are explained later in the text.

12.3 Enterprise Resource Planning (ERP)

Enterprise resource planning (ERP) is a computer application specifically designed for industrial and manufacturing units. It consists of several modules that operate in collaboration with each other and look after the overall industry. For each department, ERP contains a dedicated module. ERP is the automation of the manufacturing industry from raw material to end product and all the processes that are directly or indirectly involved in this process. Figure 12.1 shows the major modules of ERP software. ERP modules are explained as follows:

12.3.1 CAD

Computer-aided design (CAD) is frequently used for design and drafting, generating reports, three-dimensional modeling, finite element analysis, and as an input source for computer-aided manufacturing (CAM). The process of CAD includes three phases; designing the geometric model, analysis of generated model against various physical quantities, and optimization and visualization of computer graphics based on results and analysis. The advantage of CAD in the textile industry is to design and analyze the product in a shorter span of time with almost zero cost of production. It helps in all the areas of the textile process including selection of product design, visual merchandising, and product development. The CAD has revolutionized the fabric-designing from graph paper and stencils to mouse and stylus. The CAD systems are developed in

Fig. 12.1: Enterprise resource planning.

computer languages such as FORTRAN, Java, and Python. The CAD system usually comes with either an integrated module or standalone products. Examples of an integrated module of CAD systems are computer-aided engineering, finite element methods, CAM including computer numerical control (CNC), photorealistic rendering, and product data management. The standalone CAD systems include Auto-CAD, solid works, Real CAD pro, Rhino 3D, and Iron CAD. The aforementioned standalone CAD systems are especially developed for mechanical engineering. There are other CAD systems such as TexGen, CoulorTEx, Modaris, TukaCAD, Lectra, ReachCAD, Opti-tex PDS, Audaces apparel, and GT Resource which are especially designed for the textile industry [2]. A few textile CAD systems are tabulated in Tab. 12.1.

12.3.2 CAM

CAM is the use of computer software to control machine tools and related machinery in the manufacturing of workpieces. The CAD technology together with CAM technology is called CAD/CAM. The main objective of CAM is to improve productivity and efficiency. The application of CAM ensures results are consistent with high accuracy on a large production scale. The CAM is frequently used to store textiles designs for repeated printing orders. Various types of CAM machines have been developed for serving the textile industry. These machines are based on CNC programming. In textile processing, these machines are used to pick up the fabric from the store, spread and

Tab. 12.1: Few textile-based CAD systems for weaving and fashion designing.

S. no.	CAD system	Functions
1	TexGen	Specialized for two-dimensional and three-dimensional weave design [3, 4]
2	Modaris	Pattern design system [5]
3	JacqCAD	Jacquard designing, editing, creating loom control files, and punching of textile designs [6]
4	Textronic	Dobby design, jacquard design, Carpet design, 3D design, draft, and peg plan [7]
5	DigiFab	Digital textile printing [8]
6	AVA Weave	Checks, stripes, and dobby or jacquard designs, including yarn simulations and multiple color ways [7]
7	Scotweave	Technical textiles for industrial, commercial, and geo-textiles; dobby or jacquard designs. Yarn design and cross-section view, 3D weave, and 3D visualization tool to see ScotWeave fabrics realistically draped over images of garments and other real-life objects [9]
8	Arahne	Dobby and jacquard designs, color management, fabric price calculations, yarn consumption calculations, and 3D fabric simulation [10]
9	WiseTex	Weave geometry [11]
10	DesignScope Victor	Dobby and jacquard designs, three-dimensional weave designs [12]
11	Textile Vision	Color management, two-dimensional weave designs, yarn consumption calculations, and frame plan [13]
12	Optitex	Virtual prototyping; 2D CAD/CAM patterning and fashion designing [14]

cut the fabrics, label and transport the cut fabric pieces for assembly, and move the cut fabric pieces around the factory on an overhead conveyer. These machines are also used for automatic buttonholing and automatic embroidery.

The CAM is frequently used in dobby and jacquard designs and also to control the servo motor for positive control of let-off and take-up mechanisms. In dobby and jacquard designs, CAM is used to prepare and archive weave designs for both electronically and card-controlled dobbies [15]. With the help of CAD/CAM, achieving good quality with high accuracy in a shorter span of time has become possible. It eases the modification of pattern changing.

iTextile project [1] is a good example of CAD/CAM designing in the textile industry, which is based on an intelligent searching system for the development of a smart woven fabric database as shown in Fig. 12.2.

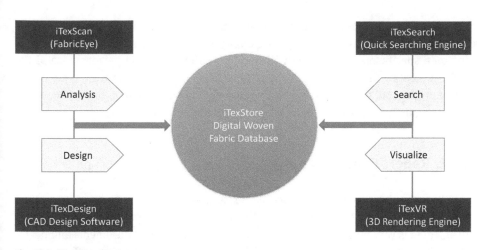

Fig. 12.2: iTextile project.

The digital scanning engine using FabricEye for this database can collect and analyze the dual-side images of fabrics for the purpose of creating image-indexing for each weave pattern style.

The quick search engine for this database, it is a specific search engine for fabric materials, which can do searching in three kinds of ways: image, color, and keyword.

The rendering engine for 3D fabric simulation is used to demonstrate and visualize the fabric design based on 3D rendering and simulation technology, and also provides the online interactive design and evaluates the fabric design directly.

All the above intelligent searching systems are connected to CAM, which helps to convert fabric databases into a product.

12.3.3 Other Areas of ERP Helping

ERP is not only helping in the design (CAD) and manufacturing (CAM) process of the product but it also looks after the overall process of the textile industry. ERP deals with almost all the departments starting from procurement, production, distribution, accounting, human resource, corporate performance and governance, customer services, and sales. It also covers the areas of business intelligence, enterprise assets management, e-commerce, and other business-related strategies.

12.3.4 Machine Monitoring

The machine monitoring module of the ERP for textiles provides the layout of the factory, indicating machines with interactive graphics, colors, symbols, and icons depicting their status. This module aids the managers to monitor the machine's efficiency and its current status (i.e., running or stopped), stoppages and their causes, and the order running on a specific machine. Along with this, it also monitors the performance of the worker operating the machine and the performance of the machine w.r.t. a particular type of product. All these features help in the instant identification of a problem, which leads to maximum throughput and increases overall productivity.

12.3.5 Report Generation

As ERP stores all the data in the central database, it allows the user to generate customized reports. These reports give a detailed insight into the key performance indicators. This module is very much customizable according to the user's requirements and needs. The key factor which users want to highlight becomes easily accessible. Different types of reports help managers to get the intuition about the business progress, and to make better decisions. These reports also help to develop and reframe the production/marketing strategies according to the latest changing trends. It is also possible to improve customer relationships by undermining customer purchase and sale patterns.

12.3.6 Production Plan

This module of ERP is somewhat like a planning board, with an extensive graphical interface. It integrates with the machine monitoring module and a central database, giving a view of the production plans. Its basic or fundamental functionality is to ensure the availability of the raw material required for production along with the use of production capacity at its maximum. It can calculate the time needed for an order to complete and then schedule the respective order on the relevant and suitable (highly efficient for that particular order) machine. Production orders can be added automatically or manually by the manager from the marketing module. Along with this, it also allows the user to define the time and quantity of raw material required for that particular order. With the use of this module, scheduling of the production order can be done on the level of individual steps involved in manufacturing. Reliable delivery dates can be communicated with customers as it schedules the target dates for each production step.

12.3.7 Inventory Management

Inventory management deals with the management of the inventory of all the resources and the raw materials like fiber, yarn, grey fabric, apparel, chemicals and dyes, and finished products. It also deals with the work-in-progress (WIP) and production stocks. Monitoring and providing the technical details about the product like piece length, width, weight, and defects of the fabric. Product and materials are reported as the product travels through the production process. Average price, shelf life, expiration date, quality grade, and original batch are also maintained by this module.

12.3.8 Maintenance

This module keeps track and gives a maintenance overview of the machines. It schedules the machine maintenance according to the production plan. It keeps track and monitors maintenance activities; lubrication of machine parts and replacement of faulty parts are required on which machines. It also maintains spare parts and accessories store rooms. Generation of reports on the maintenance activities, the machine broke downs, and spare parts are also possible.

12.3.9 Machine Setting

The electronic and mechanical settings of the machine are recorded in this module. Settings are stored with respect to product barcodes, which can be retrieved and reapplied. Weave Assist System introduced by Toyota Textile Machinery, Inc. is an example of such automation. This gives the flexibility to reapply the most promising settings for a specific product to obtain the optimal result and efficiency.

12.3.10 Quality Management

The major task of the quality management module is an assurance of required product quality as demanded by the customer. This module monitors and control incoming goods, quality control during the production process, and testing in the laboratory. Textile product is marked against their quality and the product history is maintained in this module. It also ensures quality by generating timely reports. Also provides tracking in case of customer claims and provides a graphical display of test results.

12.4 Image Analysis

Digital image analysis is the most common technique, used to identify the weave and color repeat of the fabric. A high-resolution image of woven fabric, converted to digital data, is used for the recognition of the weave pattern. This digital data is transmitted and reflected, to identify the warp and weft overlaps, the number of yarns, size, and color by comparing them with grayscale values.

12.5 Data Mining

Data mining is an interdisciplinary subject of information technology, driving its roots in data management, artificial intelligence (AI), machine learning (ML), and statistics. It is also known as the knowledge discovery from data, which is a process of extracting useful information from data. Generally, the steps involved in data mining are shown in Fig. 12.3.

With time, technology has progressed at a very fast rate. This erupting progress, computerization of society along with cheaper, powerful storage devices makes it possible for the data to be stored in huge amounts of terabytes and petabytes. Each and every field of life starting from business society, banking, finance, science, engineering, and medical sciences are getting benefits from data mining. All of these fields are the sources for generating massive datasets, which include sales transactions, stock-trading records, product descriptions, sales promotions, company profiles and performance, and customer feedback. But the problem is humans are incapable of analyzing this amount of data. This problem gave rise to the birth of an emerging and most rapidly growing field, "Data Mining." Data mining not only analyzes the data but also presents useful knowledge, extracted from the data in some presentable and interpretable format that can be easily comprehended by humans. Different techniques involving AI and ML algorithms have been developed, that enable the computers to automatically explore the dataset, and analyze and extract the information useful to our problem solution.

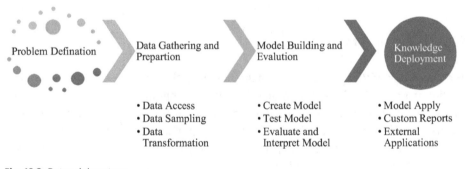

Fig. 12.3: Data mining steps.

12.5.1 Machine Learning and Artificial Intelligence

ML and AI have spread their applications in various fields of textiles. Predictive models based on ML algorithms can predict yarn crimp with up to 98% accuracy which is used to calculate yarn consumption during weaving and model the mechanical properties of a fabric [16]. Similarly, AI is being successfully used to evaluate the production planning, sewing thread consumption, seam strength, elongation at break, performance of a textile product, and estimate the several properties of the textiles like yarn hairiness, air permeability, thermal insulation, pilling, sensorial comfort, and moisture and heat transfer rate. ML and AI also help in approximating color mixing for textile printing and dyeing. An artificial neural network helps in quality control at each step of textile products from fiber, yarn, and fabric to the final garment by detecting the defects. In the same way, wrinkles, seam strength and quality in the fabric, and dimensional features like the fit of the garment and sizing system for ready-made apparel, can also be detected by these tools. Different performance functions such as drape, fabric hand, wicking, and cut and tear resistance are also analyzed to optimize the overall functionality of a textile product.

The utility of ML and AI is not only limited to the manufacturing process of the textiles but they are also being used for the analyses of marketing approach, sales behavior, consumer attitude, global market penetration, and export parameters. The COVID pandemic has changed the ways of business and consumer conduct toward buying, and a major shift has occurred in online marketplaces such as Amazon, eBay, and Ali express. AI tools are at the forefront to understand individual consumer needs and provide means to cater to them.

12.6 Blockchain and Suitability

Blockchain technology is a decentralized distributed ledger that keeps the track of digital assets. The foremost feature of this technology is the security of data, which means data on the blockchain cannot be modified. This technology is commonly linked with cryptocurrency but it has its usage beyond that. In textiles, especially in the fashion industry, several brands are using end-to-end supply chain traceability to ensure sustainability in the manufacturing, retailing, and social processes.

The concept of sustainability is to use such practices which does not deplete the natural resources and keep the ecosystem enact. It is divided into three main components, i.e., environmental, economic, and social. The fashion industry is rapidly shifting toward sustainable practices but there is a major traceability gap as shown in Fig. 12.4. Textile processing is complex and comprises numerous tiers. Blockchain technology-based solutions like Arianee, Provenance, Trust Trace, TextileGensis™, everledger™, IBM Blockchain Transparent Supply (BTS), Virgo™, AURA™, and MasterCard™ are currently working with 100 major textile brands to ensure sustainable practices.

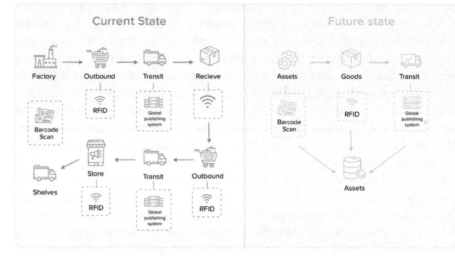

Fig. 12.4: Current and future traceability state [17].

References

[1]   T. G. Lab, iTextile. 2009 [cited 2015 July 12]; Available from: http://www4.comp.polyu.edu.hk/~ GAMA/new/itextile.htm.
[2]   A. Dwivedi, and A. Dwivedi, "Role of computer and automation in design and manufacturing for mechanical and textile industries: CAD/CAM", International Journal of Innovative Technology and Exploring Engineering (IJITEE) ISSN, vol. 3, no 3.pp. 2278–3075, 2013.
[3]   M. Sherburn, and A. Long, TexGen open source project, 2010.
[4]   H. Lin, L. P. Brown, and A. C. Long, "Modelling and simulating textile structures using TexGen", in *Advanced Materials Research*, Trans Tech Publ, 2011.
[5]   M. Stott, *Pattern Cutting for Clothing Using CAD: How to Use Lectra Modaris Pattern Cutting Software*. Elsevier Woodhead Publishing Limited, 80 High Street, Sawston, Cambridge CB22 3HJ, UK, 2012.
[6]   A. Kallivretaki, S. Vassiliadis, and C. Provatidis, Computational Modelling of Textile Structures.
[7]   P. Sinha, Computer Technology and Woven Textile Design/or CAD. 2012.
[8]   S. Gordon, *Color Management and RIP Software for Digital Textile Printing*. Kimberly Clark Corporation, iulie, 2001. p. 1.
[9]   D. Brown, M. Morgan, and R. McIlhagger, "A system for the automatic generation of solid models of woven structures", Composites Part A: Applied Science and Manufacturing, vol. 34, no. 6, pp. 511–515, 2003.
[10] A. Weave, ArahWeave® 7.1 User's manual.
[11] I. Verpoest and S. V. Lomov, "Virtual textile composites software WiseTex: Integration with micro-mechanical, permeability and structural analysis", Composites Science and Technology, vol. 65, no. 15–16, pp. 2563–2574, 2005.
[12] Company, T.D. Jacquard. [cited 2015 July 02]; Available from: http://www.designscopecompany.com/.

[13]   shinersoft. Textile Vision. [cited 2015 July 02]; Available from: http://www.shinersoft.net/textile-design.

[14]   K. Anderson, Patternmaking: Past to present. Retrieved February, 2005. 12: p. 2006.

[15]   S. Adanur, *Handbook of Weaving*. CRC press, Boca Raton, Florida 2000, pp. 135–137.

[16]   M. Z. Fazal, M. A. Abbas, Y. Nawab, and S. Younis, "Machine learning approach for prediction of crimp in cotton woven fabrics", Tehnički Vjesnik, vol. 28, no. 1, pp. 88–95, 2021.

[17]   https://laptrinhx.com/blockchain-the-technology-making-fashion-industry-transparent-1381627555/.

Muzzamal Hussain, Syed Hamza Gillani, Shahood Uz Zaman

13 Life Cycle Analysis of Textiles

Abstract: Textile is one of the most significant economic sectors in the current era, and a major portion of the population is engaged in its supply stream to a certain level. Consequently, keeping up to date on its functionality, faults, and advancements is exceptionally critical. Furthermore, with a high quantity of resource utilization and waste creation, the textile and garment sectors have significant environmental implications from the fiber manufacturing phase through the disposal point throughout the life span. This chapter presents a life cycle evaluation of the products and operations embroiled in the textile value stream, for example, textile waste reduction, regeneration, and recycling at various degrees, and critical issues involved in the life cycle methodologies. This chapter further examines the environmental consequences of the clothing and garment industries and provides the future trends, pros and cons of textile recycling, and conservation-related opinions of resources such as water, energy, and textile chemicals at various levels.

Keywords: Life cycle analysis, close loop recycling, open loop recycling, value retention, textiles

13.1 Introduction

Globally sustainable developments gained much more attention during the last decade [1, 2] due to the upsurge in population and living standards which cause an impact on the environment directly or indirectly [3]. Therefore, the development of strategies and policies is necessary to protect the environment as well as to increase the responsibility of the industrial sector to perform under these umbrellas [4].

In the global economy, textile is one of the most influential but one of the utmost polluting sectors among all the manufacturing industries [5]. The supply chain on which the textile sector is composed is very complex [6], integrated with other problems such as social and environmental [7]. The fashion industry globally generates US $1.3 trillion, along with a value chain employing about 300 million people [8].

The textile sector uses virgin raw materials and follows the traditional economic model while generating new textile products and discarding them [9]. Disposable or incinerated textiles account for 73%, of which only 1% is consumed to generate novel textiles, 14% is recycled for interior applications such as insulating or stuffing the goods, and 12% is wasted during production [8]. With existing technologies such as a linear and circular model, it is not easy to separate textiles and clothing, which are mixtures of different fibers and leads to nonrecyclable fibers.

https://doi.org/10.1515/9783110799415-013

Global trend shows that during 2000 and 2015, there was approximately a 60% annual increase in the production and consumption of textile fabric [8, 10]. However, a 36% reduction in the time of usage was observed for clothes [11]. This latter phenomenon is mainly seen as a result of the quick fashion marketing strategy. This business model mainly consists of the garments that are hitting the market in limited bunches with the newest trends of fashion and facilitates the markets continuously twice a year with fall–winter and spring–summer collections. Hence, it requires a quick response from the designers, producers, and suppliers to introduce modern garments [10]. However, adverse environmental and social effects are mainly caused by production and consumption practices.

Nevertheless, the clothes' life span and quality are reduced, which is necessary to execute the marketing strategy. For whole selling in the western markets, these outfits are mostly produced in low-income states. Consumers purchase a large number of garments triggered by fast fashion and discard them after using them only a few times. This practice leads to the generation of waste and consumes many resources and materials [12, 13].

To analyze the environmental influences of any product from "cradle to grave," life cycle assessment (LCA) is one of the most commonly used approaches in the four-step process [14]. In addition, the "cradle-to-gate" life cycle valuation of the fractional object is from resource extraction to the industry gate before it is transferred to the consumer [15], and "gate to gate" is an analysis of a portion of the material life cycle, starting with resource exploitation, and ending at the factory gate [16]. These are all different variants of LCA which have been used as a tool in various products for evaluation of the environmental impacts as well as for the sustainability.

Due to the complex textile products mentioned earlier and the associated environmental problems and social impacts in the textile sector, LCA is a conscious and standardized approach mainly discussed in this chapter. The main focus of the chapter is how textile products influence the environment by going through different procedures/ production processes and how the LCA tool can be used to assess and interpret the entire cycles of particular products, procedures/practices, and facilities, according to the International Standardization Organization (ISO) [17, 18].

13.2 LCA Methodology and Related Issues

13.2.1 Goal and Scope of the Study

The first phase outlines the study's objectives and characterizes the product to be evaluated. Control of the study goal and scope of the system to be assessed are the key challenges from the LCA perspective: what activities must be considered in the evaluation depend upon the nature and the essence of the research and the questionnaire proposed to answer. There are numerous options, some more appropriate than others, depending on the circumstances.

Attributional or consequential methods are essential considerations in the LCA perspective. The scope of the study data required for inventory is decided based on the attributional or consequential methods, for example, the attributional method is more straightforward, that is, what are the emissions of the product manufactured at this site? Whereas the powerful method mainly deals with initiated changes from the production of new products in the market on this scale. Worldwide emissions from an energy mix appropriate for the geographical extent can be employed in an ascription method, but in a substantial analysis, the highest load source of energy is usually a more appropriate selection, for example, coal for electricity. An attributional analysis might include greenhouse gas (GHG) emissions through land-use activities such as cotton production. On the other hand, emissions from indirect land-use change are significant in a consequential analysis, such as if increasing cotton farming drives other agricultural operations toward native forests.

When multifunctional processes are involved, handling the environmental allocation consequences is another general difficulty. When a single system process generates many coproducts (or services), or whenever a single input system procedure originates from an operation (or service) that created multiple coproducts, the expert must determine the number of emissions given to every flow. When a particular procedure cannot be broken down into smaller processes, the allocation may be done by segmentation or replacement, which refers to the allocation of emissions depending on weight, amounts, or market rate. If other areas of the global economy avoid emitting emissions due to coproducts, the averted releases may be deducted from the multifaceted procedure, leaving just the quantity related to the primary product. The chosen technique of allotment may significantly influence the outcome [19]. As a result, it must be done appropriately.

13.2.2 Inventory Analysis

In the second phase, inventory analysis, data from all levels of the object's life cycle, from cradle to death, is collected and analyzed. This stage computes and presents energy and raw material demands, environmental releases/emissions, and other activities for all production phases. LCA may be used on various scales, ranging from single goods to businesses and entire industries. The investigation's scope and goal determine the procedures and criteria that apply to the research. Process analysis and input–output analysis are the two main approaches to obtain LCA inventory data. However, finding meaningful inventory data is a severe difficulty in any LCA. Process analysis is a common method in the subject of LCA, and it is used in most articles. At the product level, process, or company level, it usually entails measuring any stream of energy as well as physical inputs and outputs. At the regional or national level, input–output analysis has traditionally been applied.

The process evaluation offers the benefit of particularity, but it also has the disadvantage of being time-demanding. The input–output analysis provides the benefit of requiring minimal effort from experts to incorporate information for tiny streams and reduce the possibility of resection mistakes (i.e., missing parts of the system). The drawback of the analysis is that by employing emission statistics for each economic sector, the environmental efficiency of the disastrous polluters in each sector is understated, while various environmental authorities are penalized with exaggerations. Because of the various strengths and shortcomings of the methodologies, several efforts have been made to combine them [20, 21], by employing the process examination for the essential aspects of the system beneath investigation and input–output evaluation for the facilities that assist in maintaining those networks. Both methodologies may be used to analyze textile items, generation facilities, and businesses.

The environmental variables in the frame that could be impacted by the current decision and portions of the life cycle that significantly influence the overall outcome (which can frequently be observed throughout the work owing to LCA's iterative approach) are crucial to acquiring high-quality data. For most popular industrial procedures, for example, the Eco-invent record, emission-related datasets are accessible [22–24]. When emissions come from a single specific cause and effect from a technological process, they can frequently be measured or calculated reasonably readily. However, appropriate GHG inventories or secretion aspects for certain human actions, such as varied land-use actions (e.g., agriculture and forestry), may be challenging to discover. Value chains for textile fibers and minerals may be extensive and complicated which may pose a unique problem. As a result, it may be not easy to locate relevant inventory data.

13.2.3 Life Cycle Impact Assessment of Textile

The inventory results are converted into consequences regarding ecological health, individual health, and resource deficiency in the third stage, known as impact assessment. This part of the improvement analysis may necessitate using various raw materials, modifying manufacturing techniques, or selecting one product over another. Finally, suggestions are provided depending on the outcomes of the inventory and impact phases.

The term "life cycle assessment" refers to calculating a product's entire environmental impact from conception through disposal. A life cycle exists for goods and distribution network that begins with the design of the product and resumes through raw material sourcing, manufacturing processes, transportation, usage, product reuse, recycling, and disposal. These processes influence the environment, including energy use, water use, CO_2 generation, and waste creation [25–28]. LCA is a fundamental procedure for assessing environmental impacts and identifying the most harmful phases to enhance new technologies and procedures; various publications have looked into this topic in the literature [29–33].

The life cycle analysis of textile goods includes reuse, recycling, and waste disposal as well as the recovery of basic resources, production operations (yarn and fabric), chemical finishing and treatment, clothing production, packing, and transportation. Figure 13.1 illustrates a generic model of the lifespan of a textile item [34]. Numerous inputs, including primary resources, energy, and water, as well as outputs (such as solid wastes, air, and water emissions) damage the environment throughout a textile product's life cycle. The whole life span of textile and clothing products is examined in this research, including a diverse range of steps from the creation of fiber to the end product, as well as the effects they have on the environment at different stages.

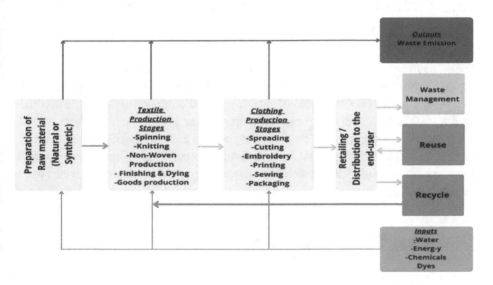

Fig. 13.1: Phases of the textile and apparel manufacturing loop/life cycle [34].

Impact assessment in the illustrated loop (Fig. 13.1) includes the issues such as carbon emissions often taking a 100-year perspective, abbreviated as global warming potential "GWP100" [35]. This choice is frequently random, which promotes the value-based viewpoint that occurs 100 years from now is not the responsibility of the decision-maker in the study. Based on different opinions in 100 years from now, we should have addressed the issue so that GHG emissions will be under control. Long-lived GHGs (those with a life span of more than 100 years) have a lesser value with a 100-year viewpoint than they would with a longer temporal perspective. A 20-year and a 500-year time horizon are two more popular options. A 20-year view may address the urgency of present climate consequences more effectively, but a 500-year perspective also addresses what will occur in the far future [35].

Finally, the environmental cost of a product or a process may seldom capture all the relevant environmental factors of a product or process decision. This necessitates the usage of numerous environmental effect classifications such as technical, cultural, and monetary studies. Due to a dearth of information, an effective impact assessment method for other impacts, or occasionally a misunderstanding of the potential threats posed by other environmental factors, and CO_2 emissions have often been the sole environmental indicators. The decision of whether or not to do an environmental impact valuation and how to supplement it with additional analyses must be driven by the choice being made.

13.3 Environmental Stewardship and Sustainability in the Textile

Today, environmentally friendly design, also known as a familiar environmental design, is critical for reducing products' adverse environmental effects. Design professionals must consider various environmental inputs and outputs when creating products with a sustainable life cycle process, including power usage, water use, toxicity, CO_2 emissions, ozone degradation, resource exhaustion, and others [36]. Designers must determine which environmental hazards (issues) to concentrate on when building a more environmentally friendly product after analyzing and comparing various product kinds [37, 38].

13.3.1 Factors Impacting Cotton Based Textile Sustainability

Sustainability refers to improvements that fulfill current generations' demands without endangering the prospective generation's ability to meet their own needs. The utilization of inherent assets, the products created, and the environmental repercussions are two critical facets of sustainability. Products or procedures that consume more exhaustible materials and harm the environment are deemed less sustainable. As previously noted, LCA has been employed to discover, quantify, and evaluate the environmental impacts (information flows) of products made from cotton fabrics. It is possible to utilize this data to evaluate textiles to different textile fibers and assess how sustainable they are [39, 40].

The World Health Organization claims that pesticide poisoning kills nearly 20,000 people in underdeveloped nations. Most of these fatalities are linked to the cotton industry [41]. Organic cotton reduces the need for fertilizers and eliminates the need for pesticides. A variety of factors influence the long-term viability of cotton-based textiles, as explained in the following discussion.

Fertility of the soil

Conventional agriculture utilizes nitrogen fertilizers in farming practices to boost soil fertility and productivity. Artificial fertilizers have negative environmental consequences; hence, sustainable cultivation, domesticated animals' dung, and lentils can be utilized to enhance soil conditions instead of artificial nitrogen composts. Animal dung, green compost, and all other options are the types of organic fertilizers which may be determined by the type and qualities of the soil [42].

Disease and pests

Plant destruction by maladies and pests accounts for 15% of production loss in cotton farming. As a result, several types of harmful pesticides are employed, resulting in major environmental consequences. To combat this, pests in organic cotton farming are handled via periodically rotating crops, a method of hybrid farming, and keeping plant density at an optimal level [43].

Water

Cotton farming uses a lot of water, which is a big problem for the environment. Irrigation is used in roughly 53% of cotton-growing areas across the world. In India, rainfall zones account for most cotton-growing areas (69%), whereas irrigation accounts for only 31% of the territory [42]. Cotton farming may be done with a variety of irrigation techniques. One of the most well-known irrigation techniques used in cotton farming is the flood-or-furrow irrigation system. Flood-or-furrow irrigation is used in around 95% of irrigated cotton fields. Portable irrigation methods are employed in a lesser percentage of fields of cotton, accounting for around 2% of all irrigated cotton fields globally. More efficient irrigation technologies should be used to increase the sustainability of cotton fiber, as the method of irrigation using furrows or floods wastes much water due to evapotranspiration into the fields.

Despite the abovementioned factors, cotton fiber manufacturing is determined to be more environmentally friendly than comparable good commercial textile fibers. Muthu et al. [44] conducted a life cycle inventory (LCI) and a life cycle impact assessment to compare cotton fiber manufacturing in terms of sustainability with that of other commercial fibers, taking into account the amount of energy and water consumed as well as the number of GHGs emitted. Based on these parameters impacting ecological sustainability, an environmental effect index (EI) was calculated using a grading scheme. EI was then utilized to compute various fiber ecological sustainability indices (ESI). Figure 13.2 depicts the estimated values of EI and ESI for various fibers [45].

It is obvious that except for flax fibers, cotton fiber has a lesser EI and a greater ESI than the majority of other fibers. Organic or natural cotton has the smallest EI and the best environmental sustainability. Flex fibers have extremely low EI and higher ESI values because they consume the least quantity of energy of all examined fibers and require perhaps less water than that cotton. Organic cotton fiber probably

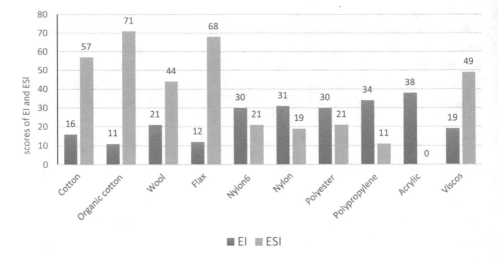

Fig. 13.2: Textile fiber EI (environmental effect index) and ESI (ecological sustainability index) values [45].

has the highest environmental sustainability due to its exclusion of synthetic pesticides and fertilizers. Although it uses less water than traditional cotton, owing to the usage of numerous chemical substances in viscose, greater EI causes more risk to human health in the viscose manufacturing phase, ecosystem conditions, the usage of significantly greater resources, and inferior ability to plants to uptake CO_2.

Furthermore, appropriate usage of by-products can increase the sustainability of cotton fiber production. Cotton fiber by-products include linter, cotton seed, and stalks. Oil may be extracted from cotton seeds. Cotton stalks may be used to create a variety of products. Research on manufacturing composites from cotton stalks is being considered [46].

Nonetheless, as previously discussed, textile products such as cotton fibers have a substantial environmental cost and worse sustainability (especially when considered with other big commercial fibers like nylon, polyester, elastane, and acrylic), due to the involvement of the production, usage, and end-of-life phases. Although using thicker cotton yarns is one essential way to lessen the environmental effect of the production process [47], on the other hand, the consequences of the usage phase can be mitigated by using more energy and water-saving laundry and drying equipment. However, there are many questions surrounding the usage and end-of-life phases of the items due to a lack of accessible data, and further research is required to get a more accurate conclusion.

13.4 Preparation and Extraction of Raw Materials

Preparing raw materials is the first step in the textile supply chain. Natural (cellulose or animal) fibers and synthetic fibers are the two primary sources of textile fibers. Traditional and organic cotton (such as rayon, linen, hemp, jute, ramie, and sisal) are natural cellulosic fibers. Cotton is used to make 40% of all textiles globally [48]. Chemically synthesized fertilizers and insecticides utilized during cultivation are the foremost significant environmental problems in cellulosic fiber manufacturing, particularly in cotton fiber. Excessive water use is still the second source of worry [28, 49].

Cotton is the most commonly employed natural fiber, among others, often utilized on the planet. Cotton cultivation in the world's agricultural land takes up about 3%, while it consumes 16% of the world's pesticides [28, 48]. Furthermore, the consumption of artificial fertilizers, pest control chemicals, technology, and power generation harms human health and the environment. Cotton cultivation also needs 7–29 tons of water for every kilogram of raw cotton fibers [50]. Hemp and flax are two more cellulosic fibers that might be regarded as the most critical sustainable fibers in the non-cotton natural fiber market [28, 51].

Fibers like wool, silk, mohair, cashmere, angora, and alpaca are examples of animal fibers. Fertilizers containing chemicals and insecticides are also employed in producing animal fiber. Furthermore, wastewater emitted during the degumming and cleaning procedures always contains certain chemicals and detergents, resulting in contamination. Furthermore, natural fibers like wool decompose and release GHG methane [48].

Organic and inorganic materials are used to make artificial fiber goods. Revived materials, altered natural components, and true synthetic polymers can be used to make synthetic fibers. Synthetic fibers may also be made using dry, wet, and melt spinning techniques [52].

Acrylic, modacrylic, polyolefin, polyester, and nylon fibers are synthetics made from synthetic polymers. Cellulose acetate, modal, and viscose rayon are semisynthetics made from natural polymers [53]. More than half of the world's garments are made from synthetic fibers. In 2011, 79.1 million tons of textile fibers were created, with 61.3% synthetic, 31.2% cotton and other natural fibres, 6% artificial cellulosic fibers, and 1.5% wool fiber [54].

The most popular material is polyester, which is used as a synthetic fiber in the world. Synthetic fibers consume less water and land in production, but they have significant environmental consequences. Synthetic fibers are typically made from nonrenewable resources (such as fossil fuels) and need much energy throughout the manufacturing process. Furthermore, increased GHG emissions are produced, and the manufacturing process necessitates using a large number of chemicals. Environmental concerns related to trash disposal, for instance, nonbiodegradability, offer a variety of health and toxicological risks, which are also negative consequences of synthetic fiber manufacturing [28, 48].

Raw materials from renewable resources and organic apparel created from organic fibers have become increasingly essential in recent years [55]. New fiber sources and renewability without agrochemicals or pesticides across the whole production procedure, from the creation of raw fiber to garment manufacturing, are key considerations in sustainable apparel [56].

13.5 Production of Textiles and Apparel

Ring spinning and open-end spinning are the two most common ways of yarn spinning. Because fibers are generated in various locations, the first stage in the yarn manufacturing process is to convey the fiber from the source to the spinning mill. Opening, carding, combing, drawing, roving, spinning, and winding are all steps in the spinning process. The primary challenges associated with environmental protection during spinning include excessive energy use, dust, fiber, and yarn waste generation [57].

Various fibers are blended during yarn spinning to achieve desired mixes, such as synthetic/natural or natural/natural fiber combinations. Furthermore, several sources such as lubricants, chemicals, water, and packaging materials are used throughout the different yarn-spinning operations. Inhalation of dust, which can induce a deadly illness termed byssinosis (also known as "brown lung"), is one of the most significant environmental concerns associated with these operations [54].

Weaving, knitting, and Non woven manufacturing are the three most common ways of obtaining fabric. During sizing before weaving, sizing agents are applied to warp threads, which can be organic or artificial, such as carboxymethyl cellulose or polyvinyl alcohol, inside the manufacturing of textile materials. All knitted and weaved fresh textiles are cleaned with water and detergents to eliminate contaminants and equipment lubricants. In addition, elevated concentrations of power usage, solid trash creation, different compounds, water, packaging materials, improper disposal, and unnecessary loudness can all be liabilities in all sorts of fabric creation activities [50, 58, 59].

Wet fabric processing consists of pretreatment, dyeing, and finishing phases. Textile finishing may also be categorized into two phases: physical and chemical. Moreover, bleaching, scouring, carbonizing, neutralizing, proportioning, desizing, mercerizing, dyeing, softening, printing, as well as other optional operations such as anti shrink finish antistatic finishing, blemish diminution, water-resistant finishes, the resistance of stain and smell, fire-proofing and moth-proofing, cleaning products, and mildew preventative measures finishes are all examples of wet processes [58]. For pollution reduction in water, the environmental consequences of textile chemicals in wet textile operations should be concentrated, and energy-efficient manufacturing techniques must be established [60, 61]. Whereas the quantity of water, energy, and chemicals consumed varies depending on the fiber treated. [62]. To dye, 1 kg of fiber, around 80–100 L of water, and several other chemicals are required. Formaldehyde and azo dye-derived aromatic amines could be employed throughout the dyeing procedure, and these ingredients might induce both cancer and allergies [48].

The textile distribution network is the final phase in clothing or apparel manufacturing. Distributing, cutting, stitching, heat straining, ironing, and packing are all parts of the garment creation process. The consumption of greater levels of power, textile waste, consumable waste, refused clothing, lubricants, water, and chemicals; utilization of packaging and holding components, during transportation release of high CO_2 emissions; and the use of child workforce in the industry are all significant factors in the assessment of environmental impacts [28].

The multiple phases of manufacturing, from essential resources to completed textile items, are typically spread throughout the globe, making logistical operations a vital element of the textile loop of supply. Following the manufacturing phase, textile and apparel items are delivered to wholesalers, merchants, warehouses, and stores by various modes of transportation, including land, sea, rail, and air. All of them have varying degrees of energy usage and CO_2 emissions ranging from 2% to 3%, followed by freight transportation which is expected to account for almost 8% of global energy-related consumption [63–66].

Fossil fuels are the most common source of air pollution. Energy utilization is a significant factor in all phases of textile generation, and there are various investigations on the subject in previous research [30, 57, 67–70]. Every stage of the finishing process uses a restricted supply of water, which becomes contaminated with chemicals such as dyes, bleaches, defoamers, and other additives. Direct emissions of noxious gases should be forbidden to avoid air pollution. Moreover, some toxins from these chemicals are emitted into the air or entrapped via our skin through garments that possess residuals of chemical compounds utilized during production. Wastewater restoration and recycling have become a critical topic of discussion to safeguard the environment and save existing water supplies [71].

13.6 The Textiles Can be Reduced, Reused, and Recycled

There are two types of textiles and apparel wastes: preconsumer and postconsumer wastes. Production of by-products known as preconsumer waste is primarily utilized to create fresh crude ingredients for automobiles, household goods, mattresses, fine yarn, paper, and other sectors. Any clothing waste generated by end user due to damage, worn out or any other reason or is considered postconsumer textile waste, and it is helpful to recycle and reuse these textiles to save water, power, dyes, and fixing substances [72].

Lowering the limit is an effective strategy used in the commodity chain to obtain a lower consumption rate of precious resources and energy. The term "recycling" refers to efforts to recover elements from items. The notion of reusing is the utilization of undamaged pieces of discarded items in industrial processes. Product/material recuperation includes actions such as repair/refurbishment and dismantling to reclaim the product's efficiency at the end of the product life [73, 74].

Regarding postconsumer garbage, there are five main options for disposal, recycling, or reuse:

1. Making a fabric or apparel item out of recovered consumable trash like plastic cans, trash polyester threads, or textiles
2. Repurposing discarded textile and apparel goods rather than tossing them away, reducing them into fibers for sound attenuation
3. Repurposing things as secondhand garments through donation stores or textile retailers
4. Generate heat, ash, and flue gas from the waste
5. Landfill disposal

Almost all of the fabrics recovered nowadays are recycled and reused. Approximately 80% of the community in various African nations buys their clothes from used clothing stores. In 2008, the University of Copenhagen researched the environmental benefits of collecting worn garments. Reduced water use of 6,000 L and 3.6 kg CO_2 emissions are both attainable, as well as 0.3 kg fertilizer utilization and 0.2 kg pesticide consumption can also be reduced by gathering 1 kg of worn clothes [34]. The life cycle stages of recycling and reuse are critical for reducing water, energy, and basic substance utilization [36].

LCA may be employed to undertake sustainability evaluations of textiles and garments during the producing phase, from primary component generation to discarded products. These figures are then evaluated to the parameters of predetermined eco-labels and specifications (such as the EU eco-mark and the Europe Oko-Tex standard). There are several kinds of accreditation systems available today, and many sustainability labels have created a wide variety of standards for textile and apparel items. Specific environmental labels, such as ISO, EU eco-mark, and Nordic sustainability labels, are solely based on the product's LCA and assess environmental consequences throughout the product life cycle, from raw product manufacturing through waste disposals [31, 59].

When a fabric approaches the limit of its usable life, the fiber is recovered for use in other items, resulting in the start of the cycle of new goods [75]. After being recycled, the initial source of material might be further categorized based on the fresh stream of products it joins. Open-loop recycling (OLR) and closed-loop recycling (CLR) are the two classes.

13.6.1 Recycling in an Open Loop

Recycling in an open-loop process is done when a product's basic materials are separated and reused in a different, sometimes unassociated, production chain (Fig. 13.3). The secondary component, in most cases, will hardly be recovered and discarded at the termination of its useful life [75]. As a result, OLR (open loop recycling) is worthwhile recycling for the initial resources of the virgin product, reducing the demand for virgin elements in the secondary product's creation.

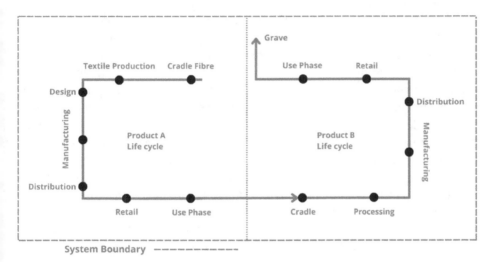

Fig. 13.3: Open-loop recycling process [75].

OLR, on the other hand, usually merely slows down a material's eventual transfer to municipal solid waste since there are constraints to how many times a substance can be recovered without losing quality.

In textiles and garments, OLR materials often comprise of

fabric wastage from the production line, such as leftovers;

clothes made from the postconsumer waste of textiles; and

polyethylene bottles from the postconsumer sector that might be converted into RPET fibers.

13.6.1.1 Description of LCA in the Framework of OLR

Solid indiscriminate dumping reduces waste from recycling in OLR from one production process into feedstock for the other. The OLR of PET bottles has been subjected to life cycle analyses, which show that using recovered PET bottles in garment fiber has energy advantages. Shen et al. [76] investigated the chemical and mechanical PET recovery and concluded that regenerated PET had more environmental advantages than virgin plastics in eight to nine areas, including power utilization, GHG emissions, acidification, and eutrophication.

In sectors that require fiber, utilizing recycled plastic and recovered yarns has obvious economic advantages. Recycled fibers may use lower energy, produce less waste, and minimize the requirement for raw resources compared to virgin fibers.

OLR of preconsumer trash is also beneficial to the environment. Muthu et al. [44] found that recovering preconsumer recyclables, also known as waste from the process, will directly reduce the product's carbon emissions. When recovered, unfortunately, not all textiles offer the same environmental advantages. Muthu et al. [77] discovered that polypropylene and polyester are the most recyclable materials, whereas nylon 6,6 is the least recyclable.

It is crucial to remember that it might be challenging to utilize LCA methods to calculate the advantages of OLR. The most important of these is determining the assessment's criteria. For instance, when preconsumer fabric waste (product life cycle A) is gathered for use in house insulation (Product System B), the LCA should take into consideration how the environmental advantages of recycling are assigned to each product system A and B [78].

As Muthu et al. [44] pointed out, further research is required to quantify the significant environmental advantages of appropriately incorporating recovered elements into the latest systems. The enormous variety of textiles makes the separation of fiber challenging for adequate recovery, and the energy spent to acquire, sort, and fabricate the crappy into fresh goods further complicates the assessment.

The extensive array of options that use pre-users and even post-users' textile waste illustrates that textile recovery works, especially when reaching lower consumer flows. In the apparel industry, OLR has shown to be effective in collecting pre- and post-user textile waste for various goods and collecting old PET bottles for reprocessing into fabrics. Garments can be used as a substrate for low-value items such as carpeting, padding, or insulation and have many uses. Nevertheless, because of the various fiber kinds and colors, the crappy fibers produced are in unappealing blacks or grays, which are inappropriate for spinning into garment standard yarn.

13.6.2 Closed-Loop Recycling (CLR)

In the garment sector, there are several approaches to describe CLR standards. This portion covers three topics: recovering clothing fabric waste (pre or postconsumer), biological and technological elements, cradle-to-cradle (C2C) flows, and repurposing current apparel.

CLR pertains to regeneration techniques in which the substance reused is the exact material used to make it: "a product reaches the manufacturing cycle of the similar good again upon use" [78]. Both pre and postconsumer garments recycled mechanically may be deemed closed circle recovered within this criterion, as long as the discarded fabric or fibers re-enter the clothing manufacturing process. Figures 13.4–13.7 show different methods.

13.6.2.1 Cradle to Cradle

To the CLR, the C2C technique is a revolutionary concept in which a CLR fabric is both reusable and recoverable within the same manufacturing loop. Trash or waste is recovered and utilized again in manufacturing the same or better value goods in the

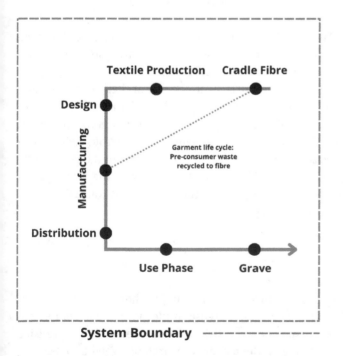

Fig. 13.4: The generation of preconsumer trash is obtained and mechanically processed for use in garments [14, 79].

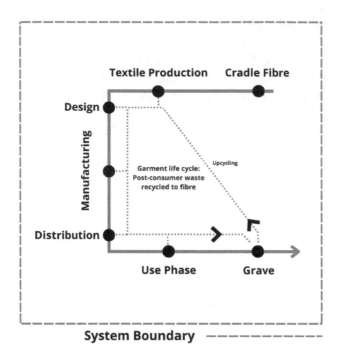

Fig. 13.5: Dismantling postconsumer waste allows the clothes to be reused [14, 79].

C2C closed concept [14, 79]. Waste is directed into one of two flows: biological or technological. Biological trash may be composted, whereas technical waste could be recycled in the manufacturing process to generate new goods.

Fibers that may be decomposed appropriately at the end of their lives to restore essential nutrients to the soil are biological CLR. Artificial goods that are not disposable are referred to as technical CLR. Synthetic polymer-based fibers like polyester, nylon, and acrylics are extensively used in textile materials. Monstrous hybrids like viscose/polyester, cotton/polyester, and cotton/spandex mixes exist in the garment industry. Mcdonough and Jaclyn Gault [79] describe the mixing of the two types of waste flows as a "monstrous hybrid," implying that the two types of waste flows cannot be efficiently separated for recycling purposes.

13.6.2.2 Closed-Loop Reuse

Reusing textiles is a similar closed-loop strategy. Reusing clothes and upcycling apparel are both connected to CLR. According to reuse in the closed loop, clothes can have numerous productive life spans on the secondary market. Although reusing clothing does not constitute recycling in the notion of separating a good into its parts, it is like CLR in that the commodity may undergo a new life cycle while still being

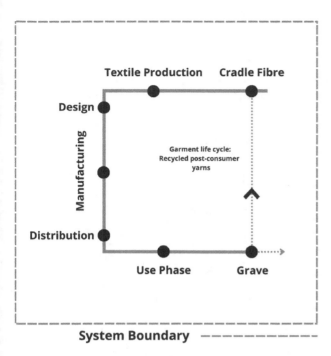

System Boundary ————————

Fig. 13.6: Postconsumer trash is discarded into fiber and reprocessed into yarn [14, 79].

produced. Hawley [80] has done a substantial study on the secondhand clothes sector worldwide. According to their observation that the secondhand clothes traded through-out the world fluctuate greatly, ranging from gently used items that are good for resale to the sought-after "diamonds" of antique or brand outfits.

13.6.3 Description of LCA Findings in the Framework of CLR

As C2C is a qualitative strategy, while the LCA is a statistical one, the sustainability advantages of employing a C2C strategy may be hard to quantify. Theoretically, the two approaches might complement each other, although that is not necessarily the case in this situation. C2C is more about creativity and anticipating system-level varia-tions rather than quantitative methodologies. According to Bakker et al. [81], the C2C strategy for CLR may not be suitable based on water and energy consumption.

The most significant consumption of water and energy is generally during the consumption stage of the clothing life span [82]. As a result, energy gains and waste avoidance via the creation of C2C terminology may not have a significant impact on critical LCA metrics like consumption of the chemical, water, and energy.

On the other hand, regeneration provides significant sustainability benefits as a CLR approach. An LCA of worn and donated apparel revealed that replacing a kilogram

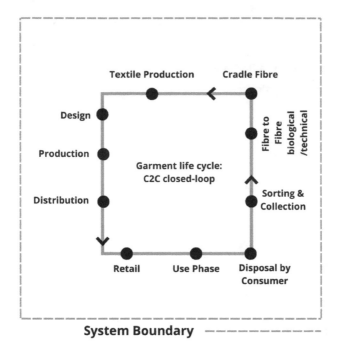

Fig. 13.7: Closed-loop regeneration from cradle to cradle [14, 79].

of virgin textiles with secondhand clothing saves 65 kWh, and replacing a kilogram of polyester with secondhand garment saves 90 kWh [72]. Considering that the usage part of the clothing life cycle consumes so much energy, clothing recovery is a realistic and crucial step toward minimizing postconsumer waste before it is recycled.

CLR has more effectiveness in garment manufacturing with the basic recovery of preconsumer waste than the postconsumer resource recovery. Because preconsumer waste, such as denim leftovers, may be reasonably neat and homogenous in fiber and color, the recovered fiber would be in a more fabulous condition for re-entering the garment distribution network. The textile scrap material can be mechanically recovered by cutting up and then respinning further into yarn, similar to the procedure mentioned earlier.

The recovered cotton fiber is often of inferior quality to fresh cotton; moreover, it can be combined with various fibers to enhance the quality. When opposed to employing virgin cotton, reusing cotton leftovers offers some advantages. According to an LCA of regenerated cotton T-shirts done for the shop Esprit, using regenerated cotton rather than new fiber saves up to 75% water [83]. Even though this assertion has not been individually proven, it is reasonable to assume that raw cotton consumes more water as compared to discarded cotton.

13.6.4 Potential of Recycled Fabrics in Future

In the realm of textile recovery, technological advancements are still being investigated. The rise of textiles in diverse fibers mixes that are challenging to distinguish for recoveries, such as polyester and cotton, is a significant impediment to successful recycling. Researchers looked at ways to extract the cotton from the polyester by dissolving the cotton with Environmental-friendly methods to salvage the polyester for recovery [84].

Another hindrance to efficient recovery is the poor color quality of fabrics. Numerous garment applications, and perhaps most broad items, are unsuitable for recycling dark gray fiber. Japanese research recommends color labeling to make it easier to turn fabrics with poor color quality into usable yarns [85]. Increased potential for CLR of textiles inside garments and continuous OLR of textiles into other goods may emerge as combined chemical and mechanical reprocessing technological advances.

Customers are more likely to demand companies that provide sustainable fabrics and examine the environmental consequences of their products due to recent environmental developments. Large businesses like H&M have had the reach to gather worn clothes on a massive scale, and by leading the way, they show that they embrace both the social and commercial benefits of recycling [86].

13.7 Conclusion

This chapter first provides and discusses the manufacturing and consumption of textiles and apparel from a global perspective. Some methodological issues in textile's LCA, for example, the study's goal, data required and collection, and environmental influence evaluations, are briefly described. This chapter also discusses the research on LCA of various textile product manufacturing phases and supply chains. Textile manufacturing is complicated with diverse material kinds, fiber creation methodology, and yarn spinning procedures; garment and apparel manufacturing techniques may be employed for various product categories. Throughout its life cycle, the textile fabrication operation uses resources such as water, fuel, and other kinds of chemicals while also producing a large quantity of waste. The primary environmental issues include pollution in the air, excessive toxicity levels, wastewater output, and solid waste generation.

Users and manufacturers are becoming increasingly environmentally conscious, seeking better environmentally favorable goods and manufacturing processes. People prefer to buy textiles created from environment-friendly resources and manufactured employing sustainable ways. This requirement will define the status of the items in the global marketplaces for the textile and garment sector.

However, numerous LCA studies are described in this chapter that is very case-sensitive and dependent on the kind of goods, utilization, reduction, reuse, recycling activities, etc., and the information accessible in the research is therefore, insufficient to make any general conclusions. It is thus critical to conduct more research to assess the environmental effect and sustainability of multiple textile goods and compare them to different sorts of goods to encourage sustainable practices.

References

[1]   Transforming Our World. "The 2030 agenda for sustainable development". in *A New Era in Global Health*, 2018.

[2]   E. Union, *Sustainable Development in the European Union. A Statistical Glance from the Viewpoint of the Unsustainable Development Goals*, vol. 43, European Union, luxembourg 2016 Doi: 10.2785/500875.

[3]   K. Fletcher, *Sustainable Fashion and Textiles: Design Journeys, Second Edition. Sustainable Fashion and Textiles: Design Journeys*, 2nd ed. pp. 1–267, 2013.

[4]   C. B. Joung, J. Carrell, P. Sarkar, and S. C. Feng, "Categorization of indicators for sustainable manufacturing." Ecological Indicators, no. 24, 2013.

[5]   S. C. Bhatia, *Pollution Control in Textile Industry*, 2017.

[6]   I. M. Sandvik, and W. Stubbs, "Circular fashion supply chain through textile-to-textile recycling," Journal of Fashion Marketing and Management, vol. 23, p. 3, 2019.

[7]   H. L. Chen, and L. D. Burns, "Environmental analysis of textile products." Clothing and Textiles Research Journal, vol. 24, no. 3, pp. 248–261, 2006.

[8]   C. Luján-Ornelas, L. P. Güereca, M. L. Franco-García, and M. Heldeweg, "A life cycle thinking approach to analyse sustainability in the textile industry." A Literature Review, vol. 12, Sustainability (Switzerland), pp. 1–19, 2020.

[9]   WCO. World Customs Organization [Internet]. 2017 [cited 2022 Jun 6]. Available from: http://www.wcoomd.org/en/topics/nomenclature/instrument-and-tools/hs-nomenclature-2017-edition/hs-nomenclature-2017-edition.aspx.

[10]   S. Yang, Y. Song, and S. Tong, *Sustainable Retailing in the Fashion Industry: A Systematic Literature Review*. vol. 9 Sustainability (Switzerland), 2017.

[11]   M. Lehmann, S. Tärnberg, T. Tochtermann, C. Chalmer, J. Eder-Hansen, and D. J. F. Seara, et al. Pulse of the Fashion Industry 2018. Bost Consult Gr Glob Fash Agenda [Internet]. p. 126, 2018. Available from: http://www.globalfashionagenda.com/pulse/

[12]   A. Moretto, L. Macchion, A. Lion, F. Caniato, P. Danese, and A. Vinelli, "Designing a roadmap towards a sustainable supply chain: A focus on the fashion industry." Journal of Cleaner Production, vol. 193, 2018.

[13]   B. Zamani, G. Sandin, and G. M. Peters, "Life cycle assessment of clothing libraries: Can collaborative consumption reduce the environmental impact of fast fashion?" Journal of Cleaner Production, vol. 162, pp. 1368–1375, 2017.

[14]   S. S. Muthu, *Handbook of Life Cycle Assessment (LCA) of Textiles and Clothing*, pp. 1–377, 2015.

[15]   A. K. Roy Choudhury, Environmental Impacts of the Textile Industry and Its Assessment Through Life Cycle Assessment. In 2014. pp. 1–39.

[16]   C. Jiménez-González, S. Kim, and M. R. Overcash, "Methodology for developing gate-to-gate Life Cycle Inventory information." International Journal of Life Cycle Assessment, vol. 5, no. 3, pp. 153–159, 2000.

[17] ISO 14044: ISO – ISO 14044:2006 – Environmental management – Life cycle assessment – Requirements and guidelines [Internet]. 2006 [cited 2021 Dec 15]. Available from: https://www.iso.org/standard/38498.html.

[18] ISO 14040: ISO – ISO 14040:2006 – Environmental management – Life cycle assessment – Principles and framework [Internet]. 2006 [cited 2021 Dec 15]. Available from: https://www.iso.org/standard/37456.html.

[19] J. Pryshlakivsky and C. Searcy, "Life Cycle Assessment as a decision-making tool: Practitioner and managerial considerations." Journal of Cleaner Production, vol. 309, 2021.

[20] L. Corominas, D. M. Byrne, J. S. Guest, A. Hospido, P. Roux, A. Shaw, et al. The application of life cycle assessment (LCA) to wastewater treatment: A best practice guide and critical review. Vol. 184, Water Research. 2020.

[21] R. H. Crawford, P. A. Bontinck, A. Stephan, T. Wiedmann, and M. Yu, "Hybrid life cycle inventory methods – A review." Journal of Cleaner Production, vol. 172, 2018.

[22] R. Hischier, B. Weidema, H.-J. Althaus, C. Bauer, G. Doka, and R. Dones, et al. Implementation of Life Cycle Impact Assessment Methods Data v2.2 (2010). ecoinvent Rep No 3 [Internet]. 2010;(3):176. Available from: https://www.ecoinvent.org/files/201007_hischier_weidema_implementation_of_lcia_methods.pdf

[23] B. Weidema, and R. Hischier ecoinvent data v2.2. the 2010 version of the most comprehensive and most popular public LCI database. 2010;3. Available from: www.ecoinvent.org.

[24] Ecoinvent Center. Ecoinvent Database Center [Internet]. 2010 [cited 2022 Jun 6]. Available from: https://ecoinvent.org/the-ecoinvent-database/.

[25] J. B. Guinée, R. Heijungs, G. Huppes, R. Kleijn, A. de Koning, and L. van Oers, et al. Life cycle assessment. Operational guide to the ISO standards. I: LCA in perspective. IIa: Guide. IIb: Operational annex. III: Scientific background. Netherlands Minist. 2002.

[26] M. Duda, and J. S. Shaw, "Life cycle assessment." Society [Internet] 1997 Nov, cited 2022 Jun 6, vol. 35, no. 1, pp. 38–43, Available from. http://link.springer.com/10.1007/s12115-997-1054-x.

[27] M. Nowack, H. Hoppe, and E. Guenther, "Review and downscaling of life cycle decision support tools for the procurement of low-value products." International Journal of Life Cycle Assessment, vol. 17, 2012.

[28] S. S. Muthu, Assessing the Environmental Impact of Textiles and the Clothing Supply Chain. Assessing the Environmental Impact of Textiles and the Clothing Supply Chain. 2014.

[29] B. Søndergård, O. E. Hansen, and J. Holm, "Ecological modernisation and institutional transformations in the Danish textile industry," Journal of Cleaner Production, vol. 12, pp. 4, 2004.

[30] V. D. W. Hmg and L. Turunen, "The environmental impacts of the production of hemp and flax textile yarn," Industrial Crops and Products, vol. 27, pp. 1, 2008.

[31] E. Nieminen, M. Linke, M. Tobler, and B. B. Vander, "EU COST Action 628: Life cycle assessment (LCA) of textile products, eco-efficiency and definition of best available technology (BAT) of textile processing," Journal of Cleaner Production, vol. 15, pp. 13–14, 2007.

[32] M. Herva, A. Franco, S. Ferreiro, A. Álvarez, and E. Roca, "An approach for the application of the Ecological Footprint as environmental indicator in the textile sector," Journal of Hazardous Materials, vol. 156, pp. 1–3, 2008.

[33] V. Thai, and A. Tokai. . . YY-J of S, 2011 undefined, *Eco-labeling Criteria for Textile Products with the Support of Textile Flows: A Case Study of the Vietnamese Textile Industry.* Thaiscience.info, p. 2, 2011.

[34] S. H. Eryuruk. "Life cycle assessment method for environmental impact evaluation and certification systems for textiles and clothing". in *Handbook of Life Cycle Assessment (LCA) of Textiles and Clothing*, pp. 125–148, 2015.

[35] J. Reap, F. Roman, S. Duncan, and B. Bras, "A survey of unresolved problems in life cycle assessment," International Journal of Life Cycle Assessment, vol. 13, pp. 5, 2008.

[36] S. E. Cepolina Textile and Clothing Industry: An Approach towards Sustainable Life Cycle Production. International Journal of Trade, Economics, and Finance 2012;

[37] M. D. Bovea and A. Gallardo, "The influence of impact assessment methods on materials selection for eco-design," Materials & Design, vol. 27, pp. 3, 2006.

[38] B. Kosińska, K. Czerwiński, and M. H. Struszczyk, "Safety and labelling requirements for textile products – Design and use aspects," Fibres & Textiles in Eastern Europe, vol. 104, pp. 2, 2014.

[39] S. Rana, S. Pichandi, S. Parveen, and R. Fangueiro Natural Plant Fibers: Production, Processing, Properties and Their Sustainability Parameters. In 2014.

[40] S. Rana, S. Pichandi, S. Parveen, and R. Fangueiro Regenerated Cellulosic Fibers and Their Implications on Sustainability. In 2014.

[41] N. A. Memon, "Organic cotton: A healthy way of life." Pakistan Textile Journal, vol. 57, no. 12, 2008.

[42] B. Prasad, and M. Dhar, Cotton Market and Sustainability in India. 2012;32. Available from: www. wwfindia.org

[43] K. Kooistra, and A. Termorshuizen, The sustainability of cotton consequences for man and environment. *Science Shop Wageningen University & Research Centre Report*. no. 223, 2006.

[44] S. S. K. Muthu, Y. Li, J. Y. Hu, and L. Ze, "Carbon footprint reduction in the textile process chain: Recycling of textile materials." Fibers & Polymers, vol. 13, no. 8, 2012.

[45] S. S. Muthu, Y. Li, J. Y. Hu, and P. Y. Mok, "Quantification of environmental impact and ecological sustainability for textile fibres." Ecological Indicators, vol. 13, no. 1, pp. 66–74, 2012.

[46] J. E. G. Van Dam, "Environmental benefits of natural fibre production and use." Proceedings of the Symposium on Natural Fibres, vol. 2009, no. 56, pp. 3–18.

[47] N. M. Van Der Velden, M. K. Patel, and J. G. Vogtländer, "LCA benchmarking study on textiles made of cotton, polyester, nylon, acryl, or elastane." International Journal of Life Cycle Assessment, vol. 19, no. 2, pp. 331–356, 2014.

[48] V. Saicheua, T. Cooper, and A. Knox, "Public understanding towards sustainable clothing and the supply chain". Nottingham Trent University, 2011.

[49] G. Dave and P. Aspegren, "Comparative toxicity of leachates from 52 textiles to Daphnia magna." Ecotoxicology & Environmental Safety, vol. 73, no. 7, 2010.

[50] E. M. Kalliala and P. Nousiainen, "Environmental profile of cotton and polyester-cotton fabrics." Autex Research Journal, vol. 1, no. 1, pp. 8–20, 1999.

[51] H. M. G. Van Der Werf, "Life Cycle Analysis of field production of fibre hemp, the effect of production practices on environmental impacts". in *Euphytica*, 2004.

[52] K. Slater, *Environmental impact of textiles: Production, processes and protection*. Elsevier, 2003. doi: 10.1533/9781855738645.

[53] H. Dayıoğlu and K. H. Elyaf Bilgisi, *Ajans Plaza Tanıtım Ve İletişim Hizmetleri Ltd. Şti, İstanbul*. 2007.

[54] S. Ellebæk Larsen, J. Hansen, H. H. Knudsen, H. Wenzel, H. F. Larsen, and F. M. Kristensen EDIPTEX – Environmental assessment of textiles (Working Report; No. 24) [Internet]. Working Report No. 24. 2007. Available from: http://www2.mst.dk/Udgiv/publications/2007/978-87-7052-515-2/pdf/978-87-7052-516-9.pdf

[55] R. Shishoo, *The global textile and clothing industry: Technological advances and future challenges* Woodhead Publishing, Cambridge, 2012. doi: 10.1533/9780857095626.

[56] URL: GC. Green Clothing: Simply A Better Way [Internet]. 2022 [cited 2022 Jun 6]. Available from: https://www.cool-organic-clothing.com/green-clothing.html

[57] C. I. Pardo Martínez, "Energy use and energy efficiency development in the German and Colombian textile industries." Energy for Sustainable Development, vol. 14, no. 2, 2010.

[58] E. Kalliala and P. Talvenmaa, "Environmental profile of textile wet processing in Finland." Journal of Cleaner Production, vol. 8, no. 2, 2000.

[59] E. Nieminen-Kalliala, "Environmental indicators of textile products for ISO (type III) environmental product declaration." Autex Research Journal, vol. 3, no. 4, 2003.

[60] S. B. Moore and L. W. Ausley, "Systems thinking and green chemistry in the textile industry: Concepts, technologies and benefits". Journal of Cleaner Production, 2004.

[61] K. Fransson and S. Molander, "Handling chemical risk information in international textile supply chains." Journal of Environmental Planning and Management, vol. 56, no. 3, 2013.

[62] S. You, S. Cheng, and H. Yan, "The impact of textile industry on China's environment." International Journal of Fashion Design, Technology and Education, vol. 2, no. 1, 2009.

[63] J. Sarkis, "A strategic decision framework for green supply chain management." Journal of Cleaner Production, vol. 11, no. 4, 2003.

[64] T. L. RcgMD, "An empirical study of green supply chain management in Indian perspective". International Journal of Applied Sciences and Engineering Research, vol. 1, no. 2, 2012.

[65] A. N. Green Logistics:, "Improving the Environmental Sustainability of Logistics." Transport Reviews, vol. 31, no. 4, 2011.

[66] S. Seuring, "Integrated chain management and supply chain management comparative analysis and illustrative cases." Journal of Cleaner Production, vol. 12, no. 8–10, 2004.

[67] H. K. Ozturk, "Energy usage and cost in textile industry: A case study for Turkey." Energy, vol. 30, no. 13, 2005.

[68] J. Brandt, K. Drechsler, and F. J. Arendts, "Mechanical performance of composites based on various three-dimensional woven-fibre preforms." Composites Science & Technology, vol. 56, no. 3, 1996.

[69] B. Van der Bruggen, E. Curcio, and E. Drioli, "Process intensification in the textile industry: The role of membrane technology." Journal of Environmental Management, vol. 73, no. 3, 2004.

[70] W. Hufenbach, L. Kroll, R. Böhm, A. Langkamp, and A. Czulak, "Development of piping elements from textile reinforced composite materials for chemical apparatus construction." Journal of Materials Processing Technology, vol. 175, no. 1–3, 2006.

[71] X. Lu, L. Liu, R. Liu, and J. Chen, "Textile wastewater reuse as an alternative water source for dyeing and finishing processes: A case study." Desalination, vol. 258, no. 1–3, pp. 229–232, 2010.

[72] A. C. Woolridge, G. D. Ward, P. S. Phillips, M. Collins, and S. Gandy, "Life cycle assessment for reuse/recycling of donated waste textiles compared to use of virgin material: An UK energy saving perspective." Resources, Conservation and Recycling, vol. 46, no. 1, 2006.

[73] S. H. Eryuruk, "Greening of the textile and clothing industry." Fibres & Textiles in Eastern Europe, vol. 95, no. 6, pp. 22–27, 2012.

[74] R. Z. Farahani, N. Asgari, and H. Davarzani, Contributions to Management Science: Supply Chain and Logistics in National, International and Governmental Environment: Concepts and Models. Supply Chain and Logistics in National, International and Governmental Environment: Concepts and Models. 2009.

[75] M. A. CurranLife Cycle Assessment Handbook: A Guide for Environmentally Sustainable Products, *Life Cycle Assessment Handbook: A Guide for Environmentally Sustainable Products*, 2012.

[76] L. Shen, E. Worrell, and M. K. Patel, "Open-loop recycling: A LCA case study of PET bottle-to-fibre recycling." Resources, Conservation and Recycling, vol. 55, no. 1, 2010.

[77] M. SS, Y. Li, H. Jy, and P. Yin Mok, Recyclability Potential Index (RPI):, "The concept and quantification of RPI for textile fibres", Ecological Indicators, vol. 18, pp. 58–62, 2012.

[78] W. Klöpffer and B. Grahl, Life Cycle Assessment (LCA): A Guide to Best Practice [Internet]. Life Cycle Assessment (LCA): A Guide to Best Practice. Wiley Blackwell; 2014 [cited 2022 Jun 8]. 1–396 p. Available from: https://onlinelibrary.wiley.com/doi/book/10.1002/9783527655625

[79] W. Mcdonough and B. M. Jaclyn Gault, Cancer, 2002.

[80] J. M. Hawley, "Digging for diamonds: A conceptual framework for understanding reclaimed textile products." Clothing and Textiles Research Journal, vol. 24, no. 3, pp. 262–275, 2006.

[81] C. A. Bakker, R. Wever, C. Teoh, and S. de Clercq, "Designing cradle-to-cradle products: A reality check." International Journal of Sustainable Engineering, vol. 3, no. 1, 2010.

[82] G. G. Smith and R. H. Barker, "Life cycle analysis of a polyester garment." Resources, Conservation and Recycling, vol. 14, no. 3–4, 1995.

[83] Esprit. Our Strategy | ESPRIT [Internet]. 2022 [cited 2022 Jun 8]. Available from: https://www.esprit.com/en/company/sustainability/towards-circularity/our-strategy

[84] R. De Silva, X. Wang, and N. Byrne, "Recycling textiles: The use of ionic liquids in the separation of cotton polyester blends." RSC Advances, vol. 4, no. 55, 2014.

[85] A. U. Motoko, K. Teruo, and S. Tetsuya, *Study on Recycling System of Waste Textiles Based on Colour.* pp. 59, 2013.

[86] A. Payne, "Open-and closed-loop recycling of textile and apparel products". in *Handbook of Life Cycle Assessment (LCA) of Textiles and Clothing*, 2015.

Anum Nosheen, Munir Ashraf, Saba Akram

14 Environmental Issues and Sustainability in Textile Industry

14.1 Introduction

The textile sector is among the largest global environmental threats with an annual production of 60 billion kilograms of fabric and water use of up to 9 trillion gallons [1]. Textile wastewater is characterized by its elevated pH, chemical oxygen demand (COD), dissolved and dispersed solids, unfixed dyes and pigments, and increased biological oxygen demand (BOD) [2]. Approximately 1.3 million tons of pigments, dyes, and finishing chemicals are being used by this industry. It is reported that different processes used in textiles such as dyeing and finishing contribute to about 20% of the world's water contamination [3]. The environment could be seriously harmed by a variety of nonbiodegradable petroleum-based pigments and dyes used in the coloration of textiles, toxic substances used as different electrolytes, dye fixers, finishing chemicals, and binders or cross-linkers used in the dyeing and finishing systems. Hazardous effects, due to the presence of these chemicals in textile wastewater, have led to various diseases such as respiratory issues and skin diseases in the laborers working in that environment [4]. There are two major divisions of textile sector:

- Dry processing: This involves the most of manufacturing and fabric assembly procedures carried out, and these processes practically do not require water such as weaving, blow room, and carding.
- Wet processing: Water is employed either in processing solutions or in their fresh state, or both such as chemical processing of textiles which includes printing, dyeing, finishing, and washing of textiles.

The processing of textiles involves several steps from gray fabric to finished form, as depicted in Fig. 14.1. During the wet processing, various processes are involved for the removal of impurities, dyeing, and finishing of textiles. Among all these processes, dyeing and finishing of textiles are most frequently employed. During dyeing and finishing, the coloration and functionalization of textile materials are performed which is costly and energy-consuming, and involves the usage of toxic chemicals. Various chemicals including N-halamine siloxanes, biopolymers, triclosan, and polyhexamethylene biguanide are extensively used in the development of textiles with antibacterial properties [5]. Various new nanoparticles which include zinc oxide, titanium dioxide, and silver are extensively employed to improve the finishing qualities of textiles. Due to the inappropriate binding of finishing chemicals and other auxiliaries, these chemicals are discharged into the environment and cause water, air, and soil contamination [6]. In the current scenario, during the processing of textiles, these wet

https://doi.org/10.1515/9783110799415-014

chemical processing systems use a substantial amount of water, colors, chemicals, and accessories. Moreover, this sector generates a huge amount of hazardous industrial effluents that are discharged into the surrounding aquatic streams, causing ecosystem damage, water contamination, climate change, energy demand, and a water scarcity. These systems are expensive, unsustainable, and not eco-friendly due to the massive amounts of water and energy consumption. The composition of the chemical reagents utilized varies from conventional inorganic or organic chemicals to different polymers and intricately manufactured organic goods. Multiple wastewater treatment procedures including biological degradation, adsorption, advanced oxidation process, coagulation, ion exchange, and membrane separation are used in the textile industry to reduce the amount of hazardous textile contaminants in the environment [7]. Despite the fact that these toxic effluents are treated, still employment of conventional processing approaches has shown to be extremely ineffective, difficult, expensive, and highly unsustainable.

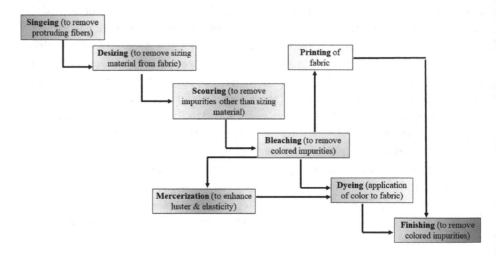

Fig. 14.1: Different steps involved the wet processing of textiles.

An overview as well as the consumption and emissions that raise environmental concerns are shown in Tab. 14.1. Any kind of toxic waste could be discharged into the environment, which includes hazardous organic substances, chlorinated solvents, nondegradable surfactants, and phosphates. These substances are the result of a variety of processes, including the preparation of fiber, dyeing or printing of fabric, bleaching, and cleaning. The workers working in printing or coloration have frequent exposure to different types of acids (such as acetic acid, formic acid, and sulfuric acid), fixatives, brighteners, and solvents, while the employees working in finishing operations are constantly exposed to flame-retardant chemicals, crease resistance chemicals, and a variety of hazardous solvents used for spotting and degreasing [8]. The effects of bleaching, dyeing, and finishing processes lead to a variety of dermatitis-related skin conditions. Exposure to intermediate dyestuffs can result in bladder

Tab. 14.1: Overview of the waste produced during textile manufacturing and processing [10].

Procedures	Source	Contaminants
Energy production	Boiler emissions	Sulfur dioxide (SO_2), oxides of nitrogen (NO_x), and particulates
Finishing	Oven emissions	Volatile organic components
Cotton-handling procedures	Emissions during the preparation and fabric manufacturing, and carding	Particulates
Sizing	Emissions due to the use of sizing materials, gums, and PVA	greenhouse gases, non-biodegradable polymers
Bleaching	Emissions due to the use of chlorine compounds	Chlorine and chlorine dioxide
Dyeing	Sulfur dyeing procedures and disperse dyeing using carriers	Aniline dyeing carriers (H_2S) and aniline vapors
Printing	Emissions	Hydrocarbons and ammonia
Finishing	Heat setting of synthetic textile materials, and resin finishing	Formaldehyde carriers, low molar-mass polymeric materials, lubricating chemicals
Chemical storage	Emissions from chemical and commodities storage containers	Volatile organic components (VOCs)
Wastewater treatment	Emissions from containers and processing vessels	Volatile organic components and toxic emissions

cancer. Bladder cancer, byssinosis, chronic bronchitis, and dermatitis are among the occupational health impacts that can affect dyers, whereas the nasal cavity diseases can affect weavers [9]. Due to poor waste management practices and a lack of awareness, these harmful solid materials and toxic chemicals are disposed of in unsafe landfills, resulting in extremely hazardous conditions for the soil, air, production of toxic waste, groundwater, and resource depletion.

14.1.1 Air Pollution

Air pollution can be described as the introduction of substances into the atmosphere that are harmful to humans or other living things, uncomfortable for them, or destructive to the ecosystem. The majority of textile industry processes emit gases into the atmosphere. The second worst pollution issue in the textile industry, following toxic effluent discharge, has been identified as gaseous pollutants. Boilers, thermopacks, and diesel generators are the primary sources of air pollution since they produce gaseous pollutants such as oxides of nitrogen gas, suspended particulate matter, and sulfur dioxide gas [11]. The major source of air pollution in the textile sector is the finishing of textiles with a variety of techniques employed to coat them. Coating materials include, for instance, lubricating oils, wetting agents, pigments, and water-repellant materials. These substances are fundamentally organic (typically hydrocarbon) compounds like solvents, oils, or waxes. The treated textiles are heated in frames, ovens, tenters, and so on to cure the coatings after they have been applied. As a result, organic compounds frequently vaporize volatile organic compounds (VOCs), often hydrocarbons [12]. Oil mists and organic gaseous pollutants are produced when textiles consisting of plasticizers, lubricating chemicals, and other substances that can be volatilized or thermally decomposed into volatile compounds are heated. Heating, tentering, calendaring, drying, and curing are the processes that can produce oil mist [13]. During several methods of spray dyeing and the carbonization of textiles, acid mist is generated. VOCs in printing paste or inks produce solvent vapors during and after solvent processing activities like dry cleaning [14]. The polycondensation of melt spinning fibers produces exhaust fumes. Processing of natural and synthetic fibers before and during spinning, napping, and carpet shearing results in the production of lint and dust.

In addition, burning fossil fuels is frequently utilized to generate the energy needed for the dyeing of textiles. The various types of air pollution during textile dyeing are driven by the consumption of energy in the textile industry (e.g., the operations of power generators and boilers). Meanwhile, emissions of different GHGs (greenhouse gases, including nitrogen, carbon monoxide, and carbon dioxide), sulfur, and carbon particles take place, potentially contributing to global warming [15]. The utilization of furnaces or boilers and electric generators, and the shipment of goods are the main causes of air pollution concerns in the dyeing sector. Textile industries pollute the air, jeopardizing human health and endangering the environment. VOC emissions are typically produced by textile industries, and the problem from these areas can be extensive and impossible to quantify [16]. The composition and the characteristics of textile effluent from different stages of the textile sector are shown in Tab. 14.2.

14.1.2 Water Pollution

Different activities carried out to meet the basic demands of human beings contaminate water resources. The bulk of abiotic components is fundamentally supported by water, which is contaminated by carcinogenic, microbiological, and organic contaminants. The contaminated water renders it unusable and unsafe for use in agriculture, industry, aquatic life, and recreational activities, in addition to having negative impacts on land, marine, and thermal contamination. The textile industry uses the most water, second to agriculture. The production of 1 kg of fabric typically results in 40–65 L of wastewater; however, this varies depending on the type of fiber being treated [17]. The total water consumption during the wet processing of textiles is shown in Fig. 14.2 [18].

Tab. 14.2: Nature and chemical composition of different textile effluents in various steps of textile sector [10].

Process	Composition	Nature
Sizing	Cellulose, waxes, polyvinyl alcohol, and starch	High BOD and COD
Desizing	Glucose, lubricants, polyvinyl alcohol, carboxymethyl cellulose, oils, and starch	Dissolved solids, high BOD, COD, suspended solids
Scouring	Wax, lubricant, fibers, sodium phosphate, sodium carbonate, sodium silicate, sodium hydroxide, surfactants	Dissolved particles, dark color, elevated pH, and COD
Bleaching	Acids, detergents, sodium phosphate, hydrogen peroxide, sodium silicate, hypochlorite, and sodium hydroxide	Alkaline and suspended solids
Mercerizing	Caustic soda	Elevated dissolved particles, low COD, and a high pH
Dyeing	Dyes, soap, mordants, acetic acid, and reducing agents	Elevated COD, lead and copper, dissolved particles, reduced suspended solids, intensely colored
Printing	Paste, mordants, soaps, starch, oils, acids, and gums	Elevated COD, suspended particles, greasy surface, and intensely colored
Finishing	Inorganic salts, hazardous chemicals, and salts	Slightly alkaline and reduced BOD

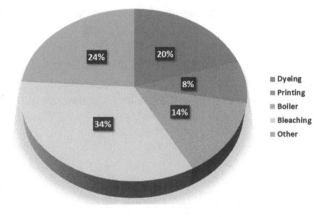

Fig. 14.2: Total water consumption in textile wet processing [18].

As reported by the prominent NGO, Greenpeace International, the processing of textiles contributes to 20% of the world's water pollution [19]. The wastewater that is produced, which comprises a wide range of hazardous and poisonous compounds employed during processing, constantly poses a threat to the environment. The primary source of pollution in the textile sector is coloring, which involves the use of colorants and other dyeing auxiliaries. Dyes that include heavy metals are particularly dangerous. The pollutants that have ecological toxicity include a wide variety of hazardous substances, including salts, detergents, formaldehyde, harmful organic compounds, soaps, biological materials, poisonous anions, emulsifying agents, and wetting agents [20]. As a result, the textile sector generates wastewater that is extremely harmful and polluting, particularly when it is combined with other chemicals and dumped without being treated. Untreated effluent from the textile dyeing sector has damaged the ability of many areas to grow food and maintain a normal way of life. It is then impossible to use the contaminated water for any other purpose because this taints the groundwater. The resulting wastewater has a high COD, elevated pH and BOD, elevated concentrations of total dissolved solids, an increased concentration of total suspended solids, and a high concentration of phenols, chlorides, and sulfates [21]. This results in serious health issues as these chemicals and colors are not easily biodegradable. Chromium and other harmful heavy metals can be found in the sludge discharged by textile treatment systems [22].

The dyeing industry is one of those wet processing methods that use a considerable amount of water. In addition to producing a significant amount of effluent, this massive water consumption also requires the use of significant amounts of chemicals, electricity, and so on. Therefore, the dyeing sector is heavily accountable for the environmental damage driven by wastewater discharge. It strongly affects the surroundings [23]:

- In general, colored water raises BOD and reduces the dissolved oxygen, which causes the death of marine species.
- The effluent generated in the dyeing sector accelerates the eutrophication-related issues.
- Some dyes contain hazardous metals.
- Toxic pollutants include alkalis, copper, lead, acids, phenols, insecticides, cyanides as well as pentachlorophenol are present in the majority of dyeing effluent.
- The toxicity of the effluent leads to a wide range of diseases, including irregular birth patterns and decreased female fertility.
- Textile effluent reduces photosynthesis.

14.1.3 Climate Change

The emission of greenhouse gases as a result of increased human activities as well as industrialization and the ensuing rise in GHG concentrations in the atmosphere are the main contributors to climate change. The textile industry is considered one of the major consumers of fuel (energy is required for steam, electric power, and transportation) and water. An estimated 20 kg of textiles is consumed year per capita, and this number continues to rise [24]. As stated by the U.S. Energy Information Administration, the textile sector ranks fifth in terms of its contribution to CO_2 emissions [25]. As a result, the textile sector is massive and one of the major emitters of greenhouse gases in the environment. Many different production procedures are needed in the textile industry. All production processes require a lot of energy for the industry. Heat is needed for some treatments, such as melt spinning, whereas refrigeration is needed for others, like dyeing, scouring, and desizing. As a consequence, energy use is crucial to several important phases of the process. According to the estimates, each meter of cloth needs between 4,500 and 5,500 kcal of thermal energy and 0.45–0.55 kWh of electrical energy [26]. This demonstrates how the textile industry contributes to climate change because of its high energy use. The textile sector uses two different types of energy sources, that is, electricity (indirect emission sources) as well as cogeneration, natural gas, and diesel fuel (direct emission sources) to meet the growing energy needs. A significant amount of generated textile effluent, which is discharged into the surrounding environment, tends to be associated with energy consumption. Energy efficiency, cogeneration, and adequate treatment of wastewater before discharge are just a few ways to reduce effluent discharge and GHG emissions [27]. Most textile sectors completely lack such measures, which results in significant pollution and changes in the climate.

The temperature of the Earth has significantly increased during the last 100 years. According to the estimates, the global temperature might increase in the range of 1.4–5.8 °C by 2100, if the current course of events is continued [28]. This is predicted to result in floods in the low-lying coastal regions, and erratic and harsh weather changes including storms, drought, and abrupt wildfires. The environment will be affected, which could lead to the extinction of some species [28].

14.1.4 Toxic Waste Generation

Waste is everything that people desire to get rid of because they no longer need it. Although the primary function of textiles is to shield the body from the elements and maintain modesty, they have evolved into indicators of personality, status, or interest in fashion. Thanks to advancements in technology, textiles are now utilized for a variety of purposes apart from just fabric production [29]. Production and consumption of textiles have become inexorable as a result of the rising global population and increased demand for new goods. The textile industries produce enormous amounts of garbage, which comprise a significant portion of municipal solid waste, from raw material procurement to the production of textiles, garment manufacturing, and transportation to retail stores.

Multiple kinds of textile solid waste materials can be categorized depending upon their nature such as packaging waste material (chemical containers), physical properties (recyclable, combustible, compostable, etc.), origin (industrial, domestic, etc.), and safety level (nonhazardous or hazardous). Production waste, preconsumer waste, and postconsumer waste are the three classifications of solid textile waste (Fig. 14.3) [30]. Production waste, which is made up of fibers, yarns, fabric remnants, and clothing cuttings produced by fiber manufacturers, processing plants, and fabric and apparel producers, is the main residual solid waste produced by the textile industries. Different waste types may be produced depending on the manufacturing processes employed, where the waste is produced. Fabric cutoffs and rolls-endings make up a significant quantity of waste, particularly in the manufacturing industry. In addition, manufacturing-related fabric defects result in production waste. Products manufactured for consumption and sale that have defects in the fabrics, the pattern, or the coloration are considered preconsumer waste. In the retail industry, unsold and damaged goods constitute preconsumer waste [30]. Any form of clothing or household item made from manufactured textiles that the user no longer requires and chooses to discard is considered postconsumer waste. When these items are worn out, broken, outgrown, or no longer in style, consumers may throw them away. Postconsumer waste is generated in immense quantities and consumed at a rate comparable to that of fibers. The majority of it is thrown away and ends up in local landfills. Compared to other waste types, the volume of postconsumer waste is relatively considerable. According to the estimates, 10.5 million tons of postconsumer textile waste are disposed of in landfills annually in the USA and 350,000 tons are disposed of annually in the UK.

Wastes from the storage and manufacturing of yarns and textiles also exist, including cardboard fabric storage reels, chemical storage barrels, and yarn cones for knitting and dyeing. Solid wastes are also produced during the processes like the manufacturing of yarn. The concept that consumers require new goods each season, promoted by the fashion industry, is one of the most significant causes of textile waste generation [31]. Due to the increase in the global population and higher living standards, there has been a significant accumulation of production and postconsumer fiber waste.

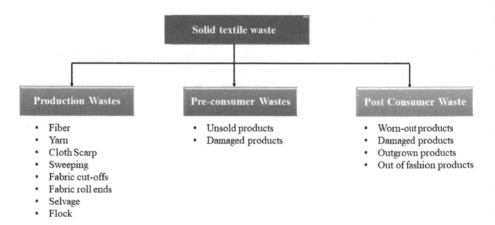

Fig. 14.3: Types of textile solid waste generated at different stages [30].

The presence of microplastics in textile wastewater is another type of toxic solid waste material [32]. Since most microplastics derived from textiles have the shape of a fiber, they are frequently referred to as microfibers. However, it should be noted that textiles consisting of natural fibers release microfibers, while other types of microplastics from textiles might appear and emerge from different kinds of accessories or materials. These materials or accessories can include patterns, buttons, glitters, and coatings.

Furthermore, microplastic fibers are released during the washing of synthetic textiles. These fibers are not collected by washing machines and end up in wastewater. Typically, these fibers are made of elastane, polyester, polyethylene, and acrylic. Thus, considering the devastating effects on aquatic organisms and human beings, removing microplastics from the aquatic system has become an urgent problem since the last 10 years [33]. Numerous aquatic ecosystems, including lakes, rivers, oceans, and urban wastewater effluents, have been shown to contain these pollutants.

14.2 What Is Sustainability

Sustainability is the potential to sustain a specific level or rate and it is the suppression of the degradation of bioresources to preserve ecological balance. Since 1987, the term "sustainability" has been widely used when the Brundtland Report of the United Nations World Commission on Environment and Development defined sustainable development as "meeting the current generation needs without jeopardizing the competence of future generations to meet their own needs." Sustainability is a miscellaneous topic comprising three main pillars: social, environmental, and economic (Fig. 14.4) [34].

14.2.1 Environmental Pillar

Among the three sustainability pillars, the environmental pillar is regarded as the most significant. Many industries are emphasizing on finding ways to lessen their social impacts, water consumption, carbon footprint, energy usage, and packaging waste. It is one of the biggest issues the world has ever faced because most of the industrial sectors in both developing and developed countries have been operating in an unsustainable manner. To overcome these challenges, we must make sure that our natural resources, including water, materials, energy fuels, and land, are used in a sustainable manner [34]. Externalities such as waste, the cost of land reclamation, and wastewater are challenging to estimate as most of the generated textile wastes directly affect the society rather than the industries.

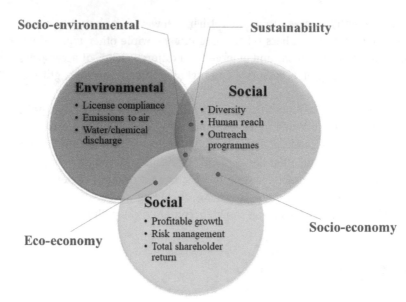

Fig. 14.4: Three different pillars of sustainability [34].

14.2.2 Social Pillar

A social pillar can be defined as the ability of a family, community, country, or organization to work at a specific living standard and harmony as well for the unlimited period that should be prolonged over time. The sustainable business relies on the community, workers, and stakeholder acceptance. Additionally, in order to further ensure the sustainability of a society, it is important to make sure that the resource usage does not proceed at a faster rate than natural resource replenishment. Waste must be released at a rate that is compatible with the amount of contamination that natural systems can withstand [35].

14.2.3 Economic Pillar

In economic progress, ecological and social aspects are highly considered. It encourages businesses and other organizations to implement the standards of sustainability beyond the requirement of legislation. Economic growth requires meeting people's needs without compromising their quality of life, especially in developing nations, while also relieving the economic burden [35].

14.3 Sustainability Legislation and Standards

Legislation and standards provide an important basis for the development and marketing of products. In the textile sector, sustainability legislation and standards are a significant tool for ensuring that the environmental effects of textile manufacturing and production are mitigated [36]. Two different types of chemicals are employed during the wet processing of textiles:

– Processing chemicals include desizing agents, surfactants, bleaching agents, and mercerizing agent
– Effect chemicals include dyes, softeners, repellents, resins, and optical brighteners

More than 72 social and environmental standards for the apparel and textile sector are included in the textile standards and legislation guide, along with a concise description of each standard's execution and significance in the textile value chain.

14.3.1 Compliance of Products

It ensures that the developed textile products comply with the setout sustainability legislation and standards such as REACH. REACH stands for Registration, Evaluation, Authorization, and Restriction of Chemicals [37]. It makes it mandatory for all chemical importers and producers to recognize and control any dangers that their products may cause to the environment and human health. In the context of textiles, REACH addresses the effect of chemicals, that is, chemicals contained within textiles [37]. Some of the basic rules of REACH are as follows;

– If chemicals are present in certain products in amounts of more than 1 ton per manufacturer or importer annually, they must be licensed for that particular application.
– Substances of very high concern (SVHC) in products must be reported if their presence exceeds 0.1% by weight threshold concentration and if more than 1 ton annually is utilized in all items.
– Suppliers of items must provide beneficiaries with enough information to permit safe use of the item if SVHC is included and surpasses 0.1% by weight, even though SVHC does not accumulate to the reporting level.

To sell the product in Europe, they must comply with REACH. Its regulation details the restriction on various hazardous materials (restricted substance list (RSL)), and preparations or articles which are produced, consumed, and placed on the EU market. The RSL is designed to provide garment and footwear manufacturers with information about the rules and legislation that restricts or outright forbids the use of various substances and chemicals in the finished home textile, clothing, and footwear items. Any substance that poses an unacceptable risk to the surrounding environment or human health during utilization could be restricted in the EU. Currently, there are 59 categories of restricted substances, involving more than 1,000 substances.

14.3.2 Compliance of Products and Production Facility

It ensures the sustainability of textile products, as well as the production of these products, that is, no toxic chemical is used during the production of textiles (processing chemicals) and the developed product does not contain any hazardous or banned chemicals (effect chemicals). REACH deals with the toxic substances and chemicals present in articles (effect chemicals). Whereas the compliance of products, as well as the production facility, involves the Zero Discharge of Hazardous Chemicals (ZDHC) standard. The previous approaches to chemical limits, which only apply to finished products, are outmoded by this standard. The toxicity of both processing chemicals and effect chemicals is addressed by ZDHC [38]. This strategy serves to protect consumers by limiting the potential effects of prohibited hazardous chemicals on manufacturing workers, nearby communities, and the surrounding environment. The ZDHC addresses environmental issues in textiles at the following three stages.

14.3.2.1 Input stage

It claims that safer products, cleaner water, and fresher air can be secured by regulating chemical inputs, which will actually benefit both people and the environment. The Manufacturing Restricted Substance List (MRSL), which contains three chapters, was designed to address all of these:
–   *Chapter 1* encompasses chemical compositions and materials used in the production and wet processing of textile materials, as well as in the manufacture and wet processing of (coated) textiles, leather, rubber, foam, and adhesives. These substances are either prohibited from use in processing or have restricted concentrations in chemicals and manufactured products.
–   *Chapter 2* suggests that ZDHC MRSL additions are capable of meeting the listing requirements outlined in the principles and procedures but lack safe alternatives at scale.
–   *Chapter 3* covers archived chemicals with high historical usage evidence but little evidence of contemporary application in industry.

14.3.2.2 Processing stage

The processing stage involves using chemicals carefully. As long as they are used properly, safer inputs can make a significant difference because processing is the vital link between inputs and outputs. This is the process to lessen their environmental burden. A management strategy offers a point of access for many supply chain stakeholders and minimizes the environmental impact. There are two parts to the ZDHC chemical management system (CMS) manual which are as follows:

- The ZDHC CMS framework, which outlines the essential elements of a CMS, that is, policy and strategic plan, is the first section.
- The ZDHC CMS Technical Industry Guide, which addresses the integration of a CMS through the supply chain, is the second section.

14.3.2.3 Output stage

The operation being undertaken with chemical inputs and processes is validated by measuring indicators including wastewater, sludge, and air quality. It assists in determining whether the output water and air are safe. The following elements are covered by the output stage:

- Wastewater guidelines
- Sampling and analysis plan
- Electronic data reporting system
- Wastewater treatment technologies
- ZDHC air emissions position paper

It does not address the whole supply chain of textile fibers including ginning, cultivation, and extraction. ZDHC aims to eradicate the use of priority toxic chemicals by emphasizing the MRSL, compliance advice, wastewater quality, and so on. The MRSL develops a list of priority chemicals and defines the recommended maximum concentration for those compounds in chemical formulations used in production facilities that process materials and cut parts for footwear and clothing.

14.3.3 Compliance of Whole Value Chain

The compliance of whole textile chain guarantees the use of safe chemicals throughout the value chain, that is, from ginning, cultivation, and extraction to the production and supply of textile products. In order to give the final consumer a reliable assurance, the global organic textile standard (GOTS), the world's leading textile standard, establishes rules to assure that textiles are organic from the time that the raw materials are harvested through environmentally and socially responsible manufacture and up to labeling. A textile item bearing the GOTS label grade "organic" denotes the presence of at least 95% certified organic fibers, while an item bearing the label grade "produced with organic" denotes the presence of at least 70% certified organic fibers [39]. The fundamental condition of GOTS is that the cultivation of fibers must be in accordance with the International Federation of Organic Agriculture Movements. The use of virgin polyester, conventional cotton, conventional angora hair fiber, acrylic, asbestos, silver and carbon fibers, and mulesed wool fibers have been restricted due to the specific fiber requirements of GOTS. The GOTS permits the use of natural materials, including organic or conventional natural fibers, bone, wood, leather, shell, and horn, as well as nonbiotic materials (such as stone, minerals, and metals), regenerated materials, and synthetic materials [39]. Besides fiber requirements, GOTS has also specified social requirements which are as follows:

- The choice of employment is voluntary
- Associational and collective bargaining rights
- No child labor will be employed
- Discrimination is not permitted
- Occupational health and safety
- No harassing or aggression
- Reimbursement and evaluation of the living earnings gap
- No risky employment is permitted
- Managing social compliance

References

[1]   A. P. Periyasamy, and J. Militky, "Sustainability in textile dyeing: Recent developments", Sustainability in Textiles and Apparel Industry, pp. 37–79, 2020.
[2]   R. Kishor, et al. "Ecotoxicological and health concerns of persistent coloring pollutants of textile industry wastewater and treatment approaches for environmental safety" Journal of Environmental Chemical Engineering, vol. 9, no. 2, pp. 105012, 2021.
[3]   R. Kant, "Textile dyeing industry an environmental hazard," 2011.
[4]   F. Parvin, S. Islam, S. I. Akm, Z. Urmy, and S. Ahmed, "A study on the solutions of environment pollutions and worker's health problems caused by textile manufacturing operations", Biomedical Journal of Scientific & Technical Research, vol. 28, no. 4, pp. 21831–21844, 2020.
[5]   L. Lin, C. Haiying, M. A.-S. Abdel-Samie, and G. Abdulla, "Common, existing and future applications of antimicrobial textile materials", in Antimicrobial Textiles from Natural Resources, Elsevier, 2021, pp. 119–163.
[6]   S. Khan, and A. Malik, "Environmental and health effects of textile industry wastewater" in Environmental Deterioration and Human Health, Springer, 2014, pp. 55–71.
[7]   S. Arslan, M. Eyvaz, E. Gürbulak, and E. Yüksel, "A review of state-of-the-art technologies in dye-containing wastewater treatment–the textile industry case", Textile Wastewater Treatment, pp. 1–29, 2016.
[8]   T. Karthik and D. Gopalakrishnan, "Environmental analysis of textile value chain: An overview", Roadmap to Sustainable Textiles and Clothing, pp. 153–188, 2014.
[9]   S. Aishwariya, and M. J. Jaisri, "Harmful Effects of Textile Wastes," 2020.
[10]   R. B. Hiremath, R. Kattumuri, B. KUMAR, V. N. Khartri, and S. S. Patil, "An integrated networking approach for a sustainable textile sector in Solapur, India", vol. 23, no. 2, pp. 140–151, 2012. 10.5379/urbani-izziv-en-2012-23-02-007.

[11] M. Yusuf, *Handbook of Textile Effluent Remediation*. New York: Jenny Stanford Publishing, 2018.
[12] B. Belaissaoui, Y. Le Moullec, and E. Favre, "Energy efficiency of a hybrid membrane/condensation process for VOC (volatile organic compounds) recovery from air: A generic approach", Energy, vol. 95, pp. 291–302, 2016.
[13] T. Meenaxi, and B. Sudha, "Air pollution in textile industry", Asian Journal of Environmental Science, vol. 8, no. 1, pp. 64–66, 2013.
[14] V. Fedele, *An Investigation of Methods to Reduce Solvent Emissions, for the Printing of Flexible Packaging Substrates*. Footscray, Australia: Victoria University of Technology, 1994.
[15] A. K. Patel, H. H. Chaudhary, K. S. Patel, and D. J. Sen, "Air pollutants all are chemical compounds hazardous to ecosystem", World Journal of Pharmaceutical Sciences, vol. 2014, pp. 729–744, 2014.
[16] R. Wu, Y. Bo, J. Li, L. Li, Y. Li, and S. Xie, "Method to establish the emission inventory of anthropogenic volatile organic compounds in China and its application in the period 2008–2012", Atmospheric Environment, vol. 127, pp. 244–254, 2016.
[17] S. Dey, and A. Islam, "A review on textile wastewater characterization in Bangladesh", Resources and Environment, vol. 5, no. 1, pp. 15–44, 2015.
[18] B. Manu and S. Chaudhari, "Anaerobic decolorisation of simulated textile wastewater containing azo dyes", Bioresource Technology, vol. 82, no. 3, pp. 225–231, 2002.
[19] A. K. Roy Choudhury, "Environmental impacts of the textile industry and its assessment through life cycle assessment" in *Roadmap to Sustainable Textiles and Clothing*, Springer, 2014, pp. 1–39.
[20] G. Crini, et al. "Advanced treatments for the removal of alkylphenols and alkylphenol polyethoxylates from wastewater", in *Emerging Contaminants*, vol. 2, Springer, 2021, pp. 305–398.
[21] A. Munnaf, M. S. Islam, T. R. Tusher, M. H. Kabir, and M. A. H. Molla, "Investigation of water quality parameters discharged from textile dyeing industries", Journal of Environmental Science and Natural Resources, vol. 7, no. 1, pp. 257–263, 2014.
[22] S. Velusamy, A. Roy, S. Sundaram, and T. Kumar Mallick, "A review on heavy metal ions and containing dyes removal through graphene oxide-based adsorption strategies for textile wastewater treatment", The Chemical Record, vol. 21, no. 7, pp. 1570–1610, 2021.
[23] N. A. Bakar, N. Othman, Z. M. Yunus, Z. Daud, N. S. Norisman, and M. H. Hisham, "Physico-chemical water quality parameters analysis on textile," in *IOP conference series: earth and environmental science*, 2020, vol. 498, no. 1, p. 12077.
[24] K. Niinimäki, G. Peters, H. Dahlbo, P. Perry, T. Rissanen, and A. Gwilt, "The environmental price of fast fashion", Nature Reviews Earth & Environment, vol. 1, no. 4, pp. 189–200, 2020.
[25] L. Peng, Y. Zhang, Y. Wang, X. Zeng, N. Peng, and A. Yu, "Energy efficiency and influencing factor analysis in the overall Chinese textile industry", Energy, vol. 93, pp. 1222–1229, 2015.
[26] A. Athalye, "Carbon footprint in textile processing", Colourage, vol. 59, no. 12, pp. 45–47, 2012.
[27] Z. Guo, Y. Sun, S.-Y. Pan, and P.-C. Chiang, "Integration of green energy and advanced energy-efficient technologies for municipal wastewater treatment plants", International Journal of Environmental Research and Public Health, vol. 16, no. 7, pp. 1282, 2019.
[28] T. Dyson, "On development, demography and climate change: The end of the world as we know it?" Population and Environment, 27, no. 2, pp. 117–149, 2005.
[29] D. J. Spencer, *Knitting Technology: A Comprehensive Handbook and Practical Guide*. vol. 16, Cambridge: CRC press, 2001.
[30] I. Yalcin-enis, M. Kucukali-ozturk, and H. Sezgin, *Risks and Management of Textile Waste*. pp. 29–53, 2019. 10.1007/978-3-319-97922-9.
[31] K. Niinimäki, and L. Hassi, "Emerging design strategies in sustainable production and consumption of textiles and clothing", Journal of Cleaner Production, vol. 19, no. 16, pp. 1876–1883, 2011.
[32] A. H. Hamidian, E. J. Ozumchelouei, F. Feizi, C. Wu, Y. Zhang, and M. Yang, "A review on the characteristics of microplastics in wastewater treatment plants: A source for toxic chemicals", Journal of Cleaner Production, vol. 295, pp. 126480, 2021.

[33] T. Poerio, E. Piacentini, and R. Mazzei, "Membrane processes for microplastic removal", Molecules, vol. 24, no. 22, pp. 4148, 2019.

[34] E. Griessler, and B. Littig, "Social sustainability: A catchword between political pragmatism and social theory", International Journal of Sustainable Development, vol. 8, no. 1/2, pp. 65–79, 2005.

[35] F. Ali-Toudert, and L. Ji, "Modeling and measuring urban sustainability in multi-criteria based systems – A challenging issue", Ecological Indicators, vol. 73, pp. 597–611, 2017.

[36] S. B. Moore, and L. W. Ausley, "Systems thinking and green chemistry in the textile industry: Concepts, technologies and benefits", Journal of Cleaner Production, vol. 12, no. 6, pp. 585–601, 2004.

[37] R. Čihák, "REACH–an overview", Interdisciplinary Toxicology, vol. 2, no. 2, p. 42, 2009.

[38] D. De Smet, D. Weydts, and M. Vanneste, "Environmentally friendly fabric finishes" in *Sustainable Apparel*, Elsevier, 2015, pp. 3–33.

[39] G. O. T. Standard, "Global organic textile standard", Recuperdo El, vol. 27, 2008.

Khubab Shaker, Yasir Nawab

15 Research Trends and Challenges in Textiles

Abstract: Textiles are used for apparel, home textiles, as well as industrial applications. This chapter discusses recent trends in textiles used for these areas. The introduction of new fibers that may be used for certain applications has also helped to introduce new textile products. In addition, the recent technological developments and automation in terms of textile machinery are also discussed in this chapter. The textile processes and their advancements are also a part of this chapter.

Keywords: research, textile machinery, textile processes

15.1 Introduction

Textiles are an essential need of a person, especially for clothing and home textiles. But the advancements in technology introduced new trends in textiles as well. The emergence of novel materials, manufacturing and process technologies, and new products have led to significant advancements in this sector. The high-performance fibers introduced between the 1950s and 1980s provided the textile industry with an enormous number of possibilities for many applications outside traditional uses in clothing and household fabrics. The advancements in fabric production techniques (weaving, knitting, and nonwoven) revolutionized the production rates of textile fabrics.

The introduction of functional features to textile materials, including breathability, liquid repellency, and antibacterial and flame-retardant capabilities, has helped to introduce new value-added products to the market. Nanotechnology, plasma technology, microencapsulation technology, and UV-curing technology are emerging technologies, making their way steadily into the textile manufacturing process. Smart textiles involving the integration of microelectronics and sensors in textiles have resulted in various intelligent textile applications.

The apparel industry has witnessed substantial improvements in garment design/ fit as well as production speeds. Some of the technological advancements include three-dimensional (3D) body scanning, systemized material flow, sophisticated cutting equipment, robotic handling of components, automation in sewing practices, and substitute fabric joining techniques [1]. Another approach is the improvisation in the process and scientific training of workers to achieve enhanced productivity.

The key driving factor behind all significant advances in textiles is technology push, market pull, and environmental considerations. The technological push includes

https://doi.org/10.1515/9783110799415-015

developments in raw materials, new fabrication and finishing techniques, and the introduction of nanotechnology. The market pull involves the market/customer demand, for example, the requirement of functional properties (hydrophobicity, flame retardancy, UV protection, antistatic behavior, self-cleaning, etc.). The environmental considerations include a focus on renewable resource-based raw materials and modifying textile processes to enhance environment friendliness. The different aspects of recent research trends in textiles are shown in Fig. 15.1.

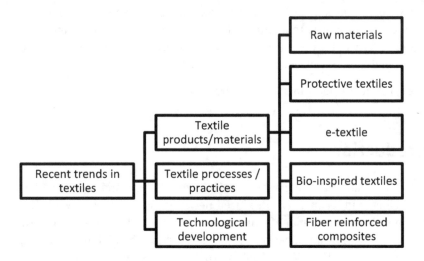

Fig. 15.1: Different aspects of recent trends in textiles.

15.2 Textile Products/Materials

The recent advancements can be classified in terms of raw materials (fibers/polymers), fabrics for protective textiles, electronics integrated into textiles, biomimetics, and fiber-reinforced composite materials.

15.2.1 Raw Materials

The innovations in raw materials, along with advancements in manufacturing technology, have been key driving forces in the textile industry. These newer raw materials are intended for producing novel materials, fulfilling both traditional and technical textile sector needs. The R&D laboratories of companies such as DuPont, Courtaulds, ICI, Teijin, and Toray have been the main source of a range of new fibers which have led to new products, applications, and markets for the textile industry.

Kelheim Fibres GmbH, a German company, has introduced 100% cellulose-based viscose fibers, making them fully biodegradable. Companies have also introduced alternate ways of producing biodegradable alternatives to man-made materials used in the textile industry. Nan Ya Plastics Corp., a Taiwan-based company, displayed a biodegradable polyester filament named Green One, which degrades into carbon dioxide, methane, and so on in approximately 98 days. Oerlikon, Switzerland, presented a new fiber production technique that allows all spinning wastes to be directed back into production. The steps involved in this technique include homogenization drying, extrusion, prefiltration, pelletizing, and downstream melt application.

New developments in traditional fibers include growing colored cotton to omit the dyeing process, organically produced, water repellent, waxed cotton for a rain suit, compatible, shrink resistance between wool and cotton fibers, non-iron or stain-resistant finishes applied to cotton and linen fibers to make more durable products. While in nontraditional textile fiber sources, it is possible to blend jute with other fibers to develop more strength. Nettle provides a fine, strong yarn that can be used for industrial purposes. Hemp is a soft but durable fiber and can be used for development purposes. Sisal has good antistatic properties and can be blended with another fiber. Pineapple and banana leaves are a source of fine silk-like fibers, but their application in textiles is still in the infancy stage. Peat fibers produce a felt-like fabric with good antistatic and absorbent attributes. Alginate is water soluble and is used for wound dressings and promotes healing.

15.2.2 Protective Textiles

These are the textiles intended to have certain properties for protection against various environmental factors. In other words, these textiles act as a barrier between the human body and the external environment [2]. For example, keeping the human body warm is a simple example of a barrier property against cold. In addition to thermal protection, there are several other desirable properties as well for high-tech materials depending on the environmental/external factor. The most common are mechanical (cut with a knife, needle pricking, bullet resistance, etc.), flame, thermal (heat or cold), chemicals (acids, alkalis, or other fluids), radiations (UV, electromagnetic, X-ray, radio-active, etc.), weather (water, wind, etc.), microbes (virus, bacteria, etc.), respiratory, and so on [3].

There has been tremendous research in these domains in the last two decades. Some of the key areas include compression and stretch-fit garments, insect-repellent textiles, fabrics with enhanced comfort properties, camouflage fabrics, impact-resistant fabrics (ballistic/stabbing/slashing/spike) [4], and flame-retardant fabrics. These advancements have been governed by technological developments in synthetic yarns, fabrication techniques, textile finishing, and nanotechnology.

15.2.3 E-textiles

Interactive textiles or e-textiles result from the integration of electrically conductive elements into textile materials. The emphasis of this domain is on the seamless integration between the fabric and the electronic elements, such as cables, microcontrollers, sensors, and actuators. There are three levels of integration of electronics in textiles [5]:

- integration of classical electronic devices (wires, integrated circuits, LEDs, batteries, etc.);
- integration of advanced electronics (conductive fibers, transistors, diodes, solar cells, etc.); and
- integration of fiber itself as a sensor or actuator.

E-textiles used in a shirt or suit could observe vital parameters such as heart rate, breathing rate, or skin temperature and could allow for the monitoring and protection of human life in dangerous situations [6]. These applications could further be extended to policemen, firemen, or soldiers. Sensors and communication technology embedded in clothing could reduce risks and help to improve emergency response capabilities by measuring environmental parameters and vital signs, as well as by warning against overstraining and external hazards. Development of smart sportswear, with integrated energy-harvesting devices, is also underway [7].

15.2.4 Biomimetic in Textiles

Biomimetics is the imitation of nature, and this category deals with textile products inspired by nature. The most famous and earliest example in this category is Velcro®, developed in 1948 by George de Mestral using polyamide fibers. The self-cleaning and water-repellent surface finish in textile materials, thermal insulation, self-healing, structural colors, and so on have been a few examples of biomimetics in textiles. The self-cleaning surface finish takes inspiration from the lotus plant and is termed as lotus effect in textiles. The leaves of lotus are superhydrophobic due to the presence of microscale bumps. It wards off water droplets, which also collect dirt from the surface while it rolls off. Air is trapped between these microbumps, which helps to create a repelling surface, resulting in superhydrophobic textiles.

Thermal insulation properties are achieved by mimicking the microstructure of a polar bear's fur (made up of tiny hollow hairs that reflect light). Butterflies make use of optical microstructures to show their colorful wings. This microstructure was mimicked to produce colorful textiles laminating 60 nano-sized layers of nylon or polyester. Swimsuits have been designed by studying the shark skin to reduce friction while swimming underwater, helping to swim faster. The penguins survive in extremely cold conditions because they can switch their skin into insulating or waterproof, by a muscle connected to the feather (a phenomenon termed adaptive insulation). An adaptive insulation jacket was produced using an ePTFE membrane along with a polyester structure (24% PTFE and 76% PE). It allows users to control the quantity of air for suitable insulation [8].

15.2.5 Fiber-Reinforced Composites

The textile fibers/yarns/fabrics also serve as a reinforcement material for fiber-reinforced polymer composites (FRPC). Fiber-reinforced plastics play an increasingly decisive role in widely varying industrial sectors due to their lightweight, good stiffness, and higher strength. The use of these lightweight materials can significantly help in the reduction of energy, as well as material demand while resulting in a considerable improvement in efficiency improvement. The aviation and automotive industry are increasingly focusing on these materials for lighter vehicles/aircraft, as well as reduced fuel consumption. The rotor blades of the wind turbine are extremely lightweight compared to the other materials. Glass, carbon, and Kevlar are the most commonly used synthetic fibers for FRPC.

The class of composite materials that emerged in the last two decades is bio-/green composites. The waste disposal, environmental concerns, and depletion of petroleum resources have fostered research in this domain of composite materials. These composites use natural fibers as a reinforcement material while biodegradable polymers are used as a matrix material. However, certain limitations involved with these materials include brittle nature, low impact strength, and tendency to absorb moisture, and are being widely researched globally. Green composites find applications in aerospace, automotive, construction, healthcare, energy, consumer goods, and other areas [9].

15.3 Textile Processes

The textile industry has transferred from a regime of extensive accumulation to a regime of intensive accumulation. The former involves growth by increased use of natural and human resources, while the latter is based on more efficient use of resources. This is achieved by the use of more sophisticated equipment, redesigning processes, minimizing wastage, enhancing productivity, and recycling waste into value-added products. The automation of the textile process owing to technological advancements is discussed in Section 15.4.

15.3.1 Approaches for Productivity Enhancement

The enhanced productivity in a garment manufacturing unit reduces the production cost and brings more profit to the business. Some of the approaches used for this purpose include:

- Conducting time and motion study to analyze time spent on each motion and correct faulty motions
- Training for operators and supervisors for capacity building and improving their management and communication skills
- Install better quality equipment and track real-time data tracking system, to acquire floor information and find the low-performing machine
- Improve line balance in a garment unit by selecting the best possible line layout, reducing setting time and scientific layout of the workstation
- Inline quality inspection at regular intervals, using traffic light system to reduce defect generation at source
- Establish a research and development team for the product

15.3.2 Recycling/Waste Management

The global fiber consumption was 120 million tons, with an average growth rate of 7%. The resources cannot sustain this growth which will result in a shortage of materials and higher prices. This price mechanism is an important factor for change. Higher fiber prices also provide incentives for more efficient processing methods (low-temperature enzymatic/catalytic scouring and bleaching of cotton, CO_2 dyeing, and digital coating), with better environmental efficiency. The higher prices also create incentives for recycling [10].

The textile industry, and specifically the dyeing, printing, and finishing industry, is responsible for a disproportionately large amount of environmental damage, and most of that damage is completely avoidable. There has to be a structured approach to reducing the negative environmental impacts of the dyeing industry. First, there is a pressing need to enforce minimum standards to reduce inexcusable pollution. Second, there has to be a drive to improve efficiency to minimize the unnecessary use of water, chemicals, and energy. Third, the industry should look to adopt specific low-impact technologies. The negative environmental impacts of the dyeing industry could be drastically reduced by applying best practice that already exists. Of course, any beneficial technological breakthroughs will be most welcome, but this chapter examines ways in which regulation and legislation can reduce environmental damage by encouraging the widespread use of existing best-available technology [11].

15.4 Technological Development

There have been significant improvements in the existing fiber spinning processes (intimate blending, core–sheath spinning, etc.), as well as new techniques have been introduced, for example, gel spinning, multicomponent melt spinning, and microfiber spinning. These techniques have enabled the production of yarns/fibers with properties more appropriate for use in technical textiles. For example, multicomponent melt spinning technology is used to produce microfibers, binder fibers, self-crimpable fibers, electroconductive fibers, and heterophil yarns. Core–sheath spinning gives yarn having the core of a certain fiber covered with a sheath of a different fiber. Thus, the properties of two different fibers may be combined in a fabric. Sensitive core materials can be protected by sheath fibers.

Advances in the manufacturing of 2D and 3D fabrics include spacer textiles, seersucker fabrics [12], simulating properties/structure of fabrics, and hydroentangled nonwovens. Spacer textiles comprise two separate fabric layers, connected by vertical spacer yarns (mostly monofilament yarn). The vertical yarns create an air space between two fabrics, providing a high degree of thermal heat insulation and improving mechanical damping. The spacer height may vary from 1 to 65 mm or more. The improved strength and durability of nonwovens make them an attractive choice for the clothing industry. The hydroentanglement processes allow the creation of new patterns, and even higher production rates may be achieved in combination with calendars. It will allow the production of 3D logos and artwork in nonwoven design with high definition. The technology named ROTIS has been developed to produce specialty nonwoven, using a toothed roller to create transversally laid layers [13].

Some of the key advancements in finishing, printing, and coating/lamination include plasma treatment for surface functionalization, digital printing, dual-side coating, hot melt coatings, and nanotechnology. The plasma treatment is an environment-friendly (water-free) process applicable to almost all types of fibers. Digital printing is a flexible and high-production rate process that can produce high-resolution, multicolor prints on textiles. It is specifically helpful for prototyping, and can produce patterns precisely according to the body geometry. Special application units for the direct coating of fabrics are integrated into the stenter entry to allow for the simultaneous coating of the top and bottom of the textile in one passage. The hotmelt and powder-coating systems have been developed due to a move away from the use of solvents in the coating industry.

Nanotechnology finds diverse opportunities in the textile industry, including the generation of nanofibers and nanocomposite fibers, and surface functionalization. The introduction of laser technology has helped to create a patterning effect by laser beam burns or melts especially in the denim industry. Laser is also used for contactless, precise, and distortion-free cutting in the garments industry. Microporous films are used in the production of breathable waterproof fabrics. These microporous membranes contain holes much smaller than the smallest raindrops, yet very much larger than a water vapor molecule, giving a good barrier against rain and at the same time resulting in good thermal comfort of the garment. The electrospun nanofiber nonwoven webs are very fragile, and the research is underway to expand their application by electrospinning continuous yarns [14].

The apparel/clothing industry is focused to modernize and automate the clothing manufacturing process. The theoretical concept of automated garment assembly consists of computerized cutting, robotic handling of cut parts, processing of clamped parts using programmable machinery, and seaming using a 3D mold with a robot-controlled sewing head. Sew-free technology (by adhesive bonding of seams), primarily for the lingerie and sportswear market, has been a widely focused domain. The other area concerns 3D body scanning and the potential for integrating scan data with 3D CAD as a way of working on the fit of garments.

15.5 Challenges for the Textile Industry

The rapid technological developments have led to high degrees of automation, resulting in faster and higher output equipment that can produce a wide range of products. However, it also poses certain challenges for the industry itself. These developments require increasing levels of investment, changing skill levels in the workforce, and faster reaction to changes in technologies or market demand anywhere in the world. Hence, technological changes impact the business model as well as all managerial functions (e.g., finance, marketing, operations, R&D, and human resource development) directly or indirectly. Textile production is highly exhaustive in terms of natural resources like raw materials, water, and energy. Therefore, evaluating sustainability in textiles has emerged as a key issue for management. Some of the main challenges facing the textile industry are:

- To be highly innovative
- To use environmentally friendly materials and processes
- To be highly efficient with high productivity
- To have a leading-edge R&D agenda
- To explore new technological borders
- To develop high-performance functional products
- To design products with high flexibility

References

[1] D. Tyler, A. Mitchell, and S. Gill, "Recent advances in garment manufacturing technology: Joining techniques, 3D body scanning and garment design", in Global Textile and Clothing Industry, pp. 131–170, Woodhead Publishing, London 2012.

[2] B. Mahltig, and T. Grethe, High-Performance and Functional Fiber Materials – A Review of Properties, Scanning Electron Microscopy SEM and Electron Dispersive Spectroscopy EDS. 2022.

[3] A. S. Richard, ed., *Textiles for Protection*. Cambridge: Woodhead Publishing Limited, 2005.

[4] K. Bilisik, *Impact-resistant Fabrics (Ballistic/stabbing/slashing/spike). Eng High-Performance Text.* Elsevier, pp. 377–434, 2018.

[5] R. Shishoo, ed., *The Global Textile and Clothing Industry*. Woodhead Publising, London 2012.

[6] F. Huang, J. Hu, and X. Yan, "Review of fiber- or yarn-based wearable resistive strain sensors: Structural design, fabrication technologies and applications", Textiles, vol. 2, pp. 81–111, 2022.

[7] V. Midha, and A. Mukhopadhyay, eds., *Recent Trends in Traditional and Technical Textiles*. ICETT 2019. Singapore: Springer, 2021.

[8] J. T. Williams, ed., *Textiles for Cold Weather Apparel*. Cambridge: Woodhead Publishing Limited, 2009.

[9] K. Shaker, and Y. Nawab, *Lignocellulosic Fibers Sustainable Biomaterials for Green Composites*. Cham, Switzerland: Springer, 2022.

[10] J. P. Juanga-Labayen, I. V. Labayen, and Q. Yuan, "A review on textile recycling practices and challenges", Textiles, vol. 2, pp. 174–188, 2022.

[11] S. Kasavan, S. Yusoff, N. C. Guan, et al. "Global trends of textile waste research from 2005 to 2020 using bibliometric analysis", Environmental Science and Pollution Research, vol. 28, pp. 44780–44794, 2021.

[12] M. Maqsood, Y. Nawab, and M. U. Javaid, et al. "Development of seersucker fabrics using single warp beam and modelling of their stretch-recovery behaviour", Journal of the Textile Institute [Internet]. Vol. 106, pp. 1154–1160, 2015. [cited 2014 Nov 12]. Available from. http://www.tandfonline.com/doi/abs/10.1080/00405000.2014.977542.

[13] J. Militký, D. Křemenáková, and M. Václavík, et al. "Textile branch and main breakthroughs of the Czech Republic in the field of textile machinery: An illustrated review", Textiles, vol. 1, pp. 466–482, 2021.

[14] S. Zainuddin, and T. Scheibel, "Continuous yarn electrospinning", Textiles, vol. 2, pp. 124–141, 2022.

Index

https://doi.org/10.1515/9783110799415-016

Printed in the USA
CPSIA information can be obtained
at www.ICGtesting.com
LVHW082036140524
780250LV00005B/738